AF808732

BIOLOGY TOPIC TEXT B: CELL STRUCTURE AND ENERGY

HAROLD M. ICKES II, M.D.

Executive Editor: Rhett C. Smith, Ph.D.

Ideal Quill Publishing Group, LLC

idealquill.com

Printed in the United States of America

10 9 8 7 6 5 4 3 2 1

ISBN 978-0615834832

Biology Topic Text B: Cell Structure and Energy

by Harold M. Ickes II, M.D.

Table of Contents

Unit 1

Diffusion and Osmosis

In this section we cover:

1) Physical and chemical gradients
2) How gradients break down living systems
3) How living systems use gradients to get things done.
4) Osmosis and its dangers to cells
5) Isotonic, Hypertonic, and Hypotonic solutions

In the previous topic text, we've talked about the basic types of molecules that are found in living systems. The next development in the discussion about the structures of life will be about cell membranes and then about the structures inside the cell. Before we can introduce that, however, it's important to understand the concepts of **diffusion** and **osmosis**. Once we have a firm grasp on these principles, it will allow us to develop a greater appreciation for the structures and importance of the cell membrane as well as other cell components.

Diffusion

The phenomenon of diffusion is simple, and a little experiment will help us outline it. Suppose you have a glass of water. Now, you put a drop of food coloring into the water. What happens? The food coloring slowly begins to spread out. Over time, it eventually spreads evenly throughout the water. This is diffusion.

Why does diffusion occur? Liquids are basically collections of molecules that are tumbling, rolling and shoving their way past each other in constant motion, like people in a dance club. The drop of food coloring is a huge clump of different molecules that are suddenly put into that highly energetic environment. That clump is said to be in high initial **concentration**,[1] meaning there is a lot of it squished into a small starting space. The water will push and shove these new molecules in random directions. Slowly over time the food coloring molecules are pushed farther and farther from each other, until finally they have all been moved at random throughout the entire liquid. They will continue to be pushed and shoved, but there is no reason to suspect that they will all be pushed randomly back into the starting place. In fact, the odds are downright astronomically remote for that to *ever* happen, unless there is some specific mechanism or machine that is purposely doing that.

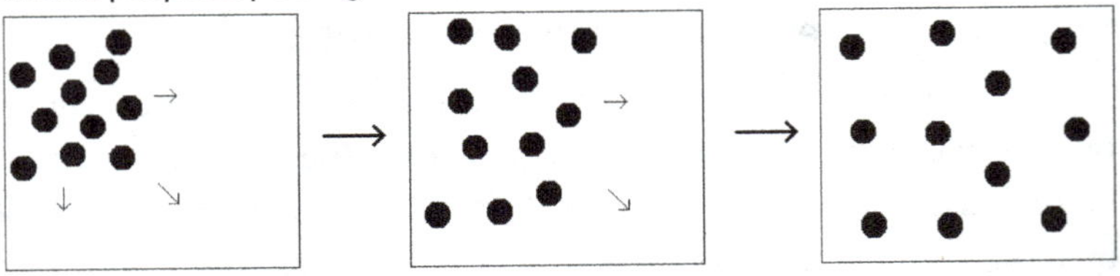

The end result of our little experiment is that the food coloring went from a high concentration (the space where it started) to low concentration (spread throughout the glass). Any time you have a difference in concentrations, there exists a **concentration gradient**. We can generalize this, in fact. A gradient refers to a difference in intensity of some physical characteristic, like concentration, pressure, temperature, or energy. Here is a more cartoonish image of what a concentration gradient might look like (high concentration at the left, low concentration at the right):

[1] Concentration is defined as the number of particles per given area of volume. A large number of particles in small volume gives a high concentration. A small number in a large volume gives a low concentration.

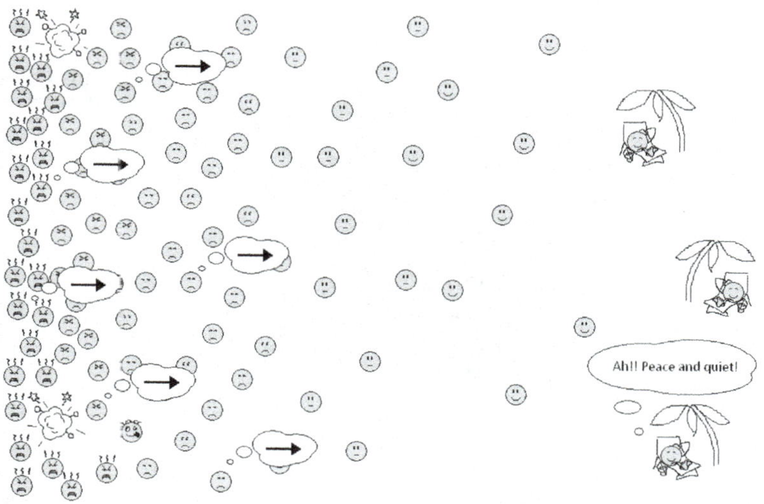

On the left, the faces are crowded and irritated, like most of my neighbors. On the right, they are spread out and relaxed. As several of the little faces indicate, there is a general tendency for the chemicals to want to go to the right.[2] That is the direction of the gradient . . . "crowded" to "spread out," or high concentration to low concentration. It's always high to low.

If we expand upon this, a **temperature gradient** might look like the next figure. Suppose a hot stone is dropped into shallow water. The heat will spread from the hot region outward toward the cold region.

---

[2] Despite my drawings, try to resist thinking of chemicals as having human ambitions. Saying that chemicals "want" to move from high to low concentration is meant as an analogy that describes a property derived from the Laws of Thermodynamics.

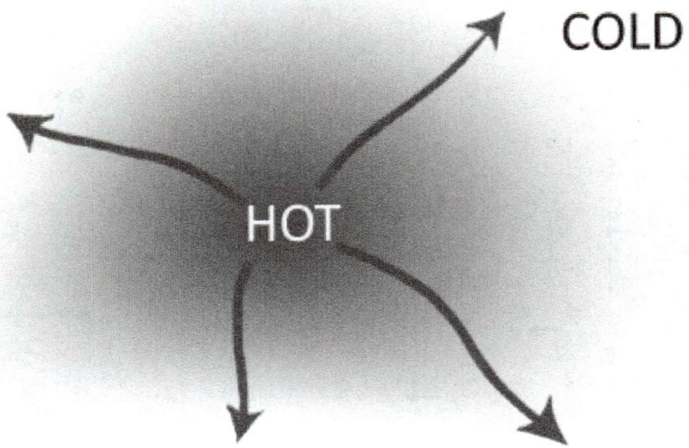

Once again, the gradient points from high to low, and eventually the heat will dissipate, and the rock will no longer be hot. When gradients exist, physical systems tend to change in a way that eliminates the gradient. In that way, it behaves just like our food coloring did. This is the natural order of things, and unless we put work into reversing the process, all gradients will slowly degrade until the temperature, the concentration, the pressure, and the energy are all spread out evenly. In that respect, pressure and energy gradients would look similar to our temperature gradient picture.

The idea of chemical and physical gradients is especially important for biological systems. Think of your body as a high concentration of all kinds of chemicals smooshed into a limited space. The natural tendency is for those chemicals to want to spread out. Tiny structures within your body are constantly breaking down and spreading out, and so without energy and materials, you quickly find yourself on the losing end of a timeless battle.

Gradients aren't always bad, however, for on a microscopic level cells sometimes use gradients to get things done. For example, the little smiley faces we just encountered all want to move toward the right. Chemicals behave in a similar fashion, and cells often put barriers (such as membranes) in the way of that movement, letting the chemicals through only after they've donated a portion of their energy. That energy is then used to get other things done.

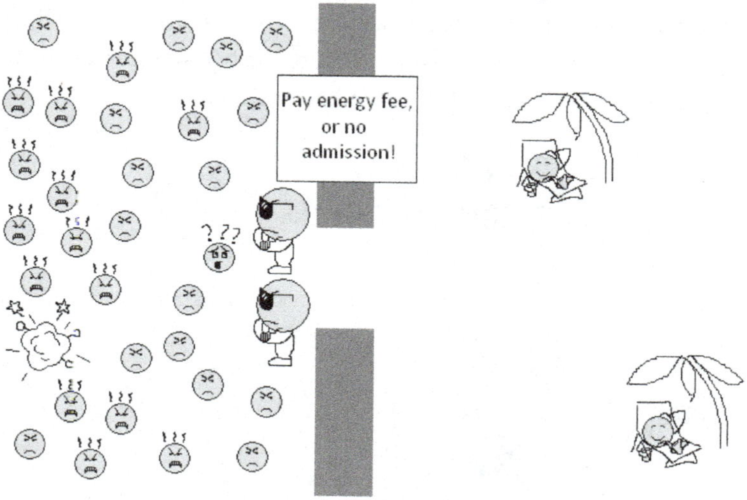

So, while gradients are always working to break down the living system, the living system can use gradients to acquire energy to get things done. In the war against gradients, we all eventually lose, but in the short term we win little battles here and there, getting a few things done in the process. All of this is due, in essence, to the basic concept of diffusion. Let's expand upon this.

Osmosis

While cells can often acquire energy from chemical gradients using membrane barriers, they must be careful when doing so. There is a potential for a dangerous process called **osmosis** (i.e. "water movement"). Suppose I put a HUGE amount of Na^+ ions inside a cell, but I don't give them a way to escape. In other words, the cell's membrane around the perimeter is a barrier that doesn't let them leave very easily. Ideally, the Na^+ will want to diffuse down their concentration gradient, (i.e., leave the cell to spread out with as much water between them as possible). They're in a pickle of a situation, because they can't leave. What happens next is rather bizarre. Water begins diffusing through the membrane.[3] The water coming into the cell is successfully diluting the concentration of Na^+ inside the cell, but it also causes the cell to expand. If too much water comes in, the membrane breaks, causing the cell to pop and die.

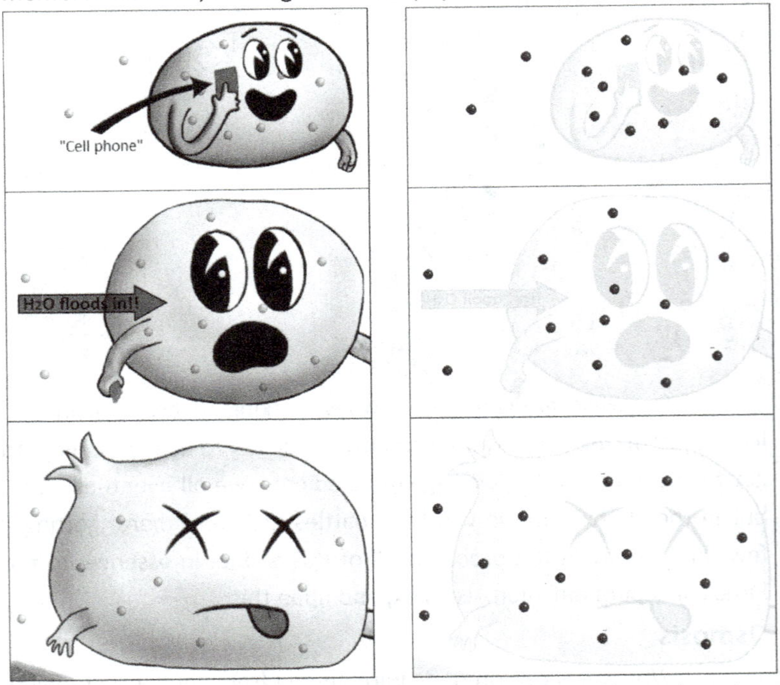

The second column in last image focuses on the sodium ions themselves. I've grayed-out the cell drawings to concentrate on the real physical phenomenon. What we see is the classic chemical

[3] As we will see in future sections, water can readily go through membranes.

gradient behavior. The sodium spreads out, and in this case it bursts the cell in the process.

The direct opposite of this can happen as well. If there is a greater concentration of Na^+ outside the cell than inside, water will leave the cell, causing the cell to shrivel. Eventually, it loses too much water and dies.

Once again, the second column focuses on the sodium ions to point out the real physical phenomenon. Diffusion makes ions want to spread out, and if cells are not careful they will get crushed or broken in the process. We regularly use this phenomenon to our advantage by packaging food with high sugar or salt content to protect it from bacteria. When the bacteria attempt to invade the food, they encounter the incredibly high concentration of sugar or salt, which causes them to shrivel and die.

Wait. First I said that cells can use membrane barriers in order to acquire energy from concentration gradients. Now it seems that having such barriers causes the cell great harm. If differences in concentration across membrane barriers can be so disastrous, how and why do cells even do it?

To answer this, let's take a closer look at osmosis. Osmosis is the movement of water. In the above scenario of the exploding or shrinking cell, we only looked at the effect of the concentration of *one* ion type in the movement of water. In actuality, the water transfer in or out of a cell depends on the *total* concentration of *all* dissolved chemicals on both sides of the membrane. Individual concentrations may differ, but as long as the combined concentration is equal inside and outside the cell, the water will stay put. For example, if there is more Na^+ outside the cell, and more K^+ inside, then the net effect is that water will not move. Think of it as Na^+ playing tug-of-war with K^+, each pulling on the water. The high outside Na^+ concentration wants to pull water out. The high inside K^+ concentration wants to pull water in. Their efforts cancel, and the water stays where it is. Neither chemical is able to spread out as it wants.

The cell finally finds an isotonic solution.

The combined concentrations are equal inside and outside the cell, so no osmosis occurs.

Another way that one might look at the phenomenon of

osmosis is through the "osmotic strength" of a solution.[4] If one solution of chemicals has the same overall concentration as another solution of chemicals, the solutions are said to be **isotonic**, ("same strength"), to each other. If one solution has a greater combined concentration than another solution, it is said to be **hypertonic** ("greater strength") to that solution. The second, weaker solution is called **hypotonic** ("less strength").

Here is a picture to round out this part of the discussion:

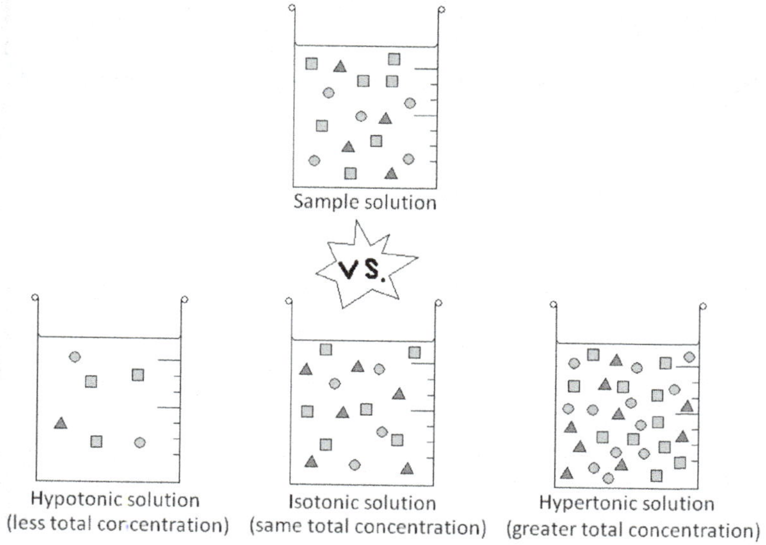

So, let's summarize a cell's behavior in terms of isotonic, hypotonic and hypertonic solutions. In an isotonic solution, a cell will have the same osmotic strength as the solution, so no water moves; the cell does not shrink or expand. In a hypotonic solution, the cell has greater osmotic strength than the "lesser strength" solution, so

---

[4] A solution is a liquid with other molecules mixed in. In the case of biology, we are usually talking about water with chemicals mixed in.

the cell pulls water into it. The cell expands and pops. In a hypertonic solution, the cell has less osmotic strength than the "greater strength" solution; the cell shrinks as water is pulled from it.

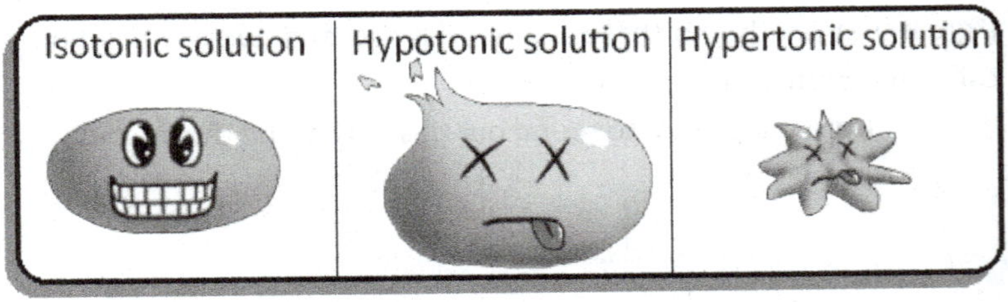

Summary

In this section, we discussed diffusion, concentration gradients, and osmosis.

- A physical or chemical gradient is an intensity difference in some measurable quantity. A system containing a gradient will tend to change in a way that slowly reduces the gradient, unless some external force such as a machine counteracts it. So:
  - Chemicals are always trying to spread out, whenever possible.
  - Heat spreads out from high temperature to low temperature, eventually leveling out the temperature.
  - Differences in pressures tend to level out.
  - Differences in energy tend to level out.
- Living systems are constantly breaking down, in part because the chemicals in them always want to spread out.
- Living systems sometimes use membrane barriers between gradients, forcing the molecules in the high concentration area to use part of their energy to get to the low concentration area. This "taxed" energy can then be used to get other things done in other parts of the cell.
- Osmosis is the movement of water from an area of low ion/molecule concentration to an area of high ion/molecule concentration. The effect is due to the combined concentrations across a barrier.
- Isotonic solutions have the same osmotic strength, so no water will move between them.
- A hypertonic solution has greater osmotic strength than some test solution (or cell) in question. Hypertonic solutions will pull water from the other solution.

- A hypotonic solution has less osmotic strength than some test solution (or cell) in question. Hypotonic solutions will have water pulled from them by the other solution.
- Cells placed in isotonic solution do not expand or shrink. Cells placed in hypertonic solutions shrink. Cells placed in hypotonic solutions expand and burst.

Unit 2

The Cell Membrane, Part 1: Structure

In this section we cover:

1) The structure and roles of cell membranes.
2) Fluid mosaic model
3) Glycoproteins
4) The three types of proteins in membranes: receptor proteins, recognition proteins, and transport proteins.
5) The Na^+-K^+ pump.

The cell is the basic unit of life. As such, the **cell membrane** (or plasma membrane) is the cell's interface with the outside environment. The cell membrane is an incredibly complex structure and serves to play three key roles: it 1) keeps the cell's contents inside the cell and everything else out, 2) selectively exchanges material across it and 3) communicates with the world outside, including other cells.

Key Roles of Plasma Membrane:
1) Isolate the cell from the outside
2) Exchange materials selectively with the outside
3) Communicate with the outside

The main molecule type in membranes is the phospholipid. Phospholipids line up one by one with their nonpolar tails snuggling up against each other in an attempt to avoid water. Their polar heads all poke into the watery surroundings. In this way, they form a sheet of phospholipids, like a crowd of people jammed into a stadium, standing room only. The nonpolar tails of the phospholipids touch the nonpolar tails of yet another layer of phospholipids whose heads point in the opposite direction, much like the same crowd standing on one massive mirror. Altogether, this arrangement is called the **lipid bilayer** (or **phospholipid bilayer**).

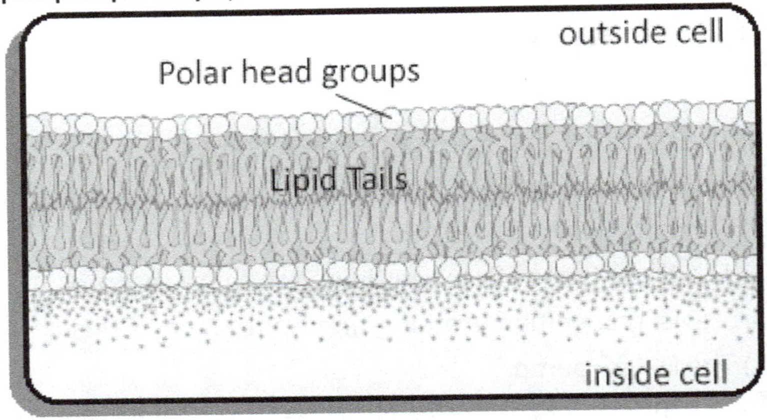

This lipid bilayer acts to isolate the contents of the cell, called the **cytoplasm**, from the water outside the cell, called the **extracellular fluid**. Ions and polar compounds are hydrophilic and usually have a hard time passing through the lipid bilayer, because it is a nonpolar barrier. The smaller the polar molecule, however, the less perfect its exclusion from the cell may be. Water, as we have seen in the case of osmosis, can pass a little at a time through the membrane, leading to potential complications for the cell.

The cell membrane has more than just phospholipids. There are also embedded proteins called **membrane proteins** that dot the cell's surface like little islands in a sea of lipids. Many membrane proteins are attached to fibrous proteins that run just under the inside surface of the cell membrane, forming a rigid skeleton that not only keeps the membrane proteins from moving, but also gives the cell its shape. There are also membrane proteins that are NOT attached to the underlying network as well. These proteins drift around like icebergs or ships on a membrane sea. This vision (i.e. membrane proteins moving about on the cell surface as if floating in a fluid) is referred to as the **fluid mosaic model**.

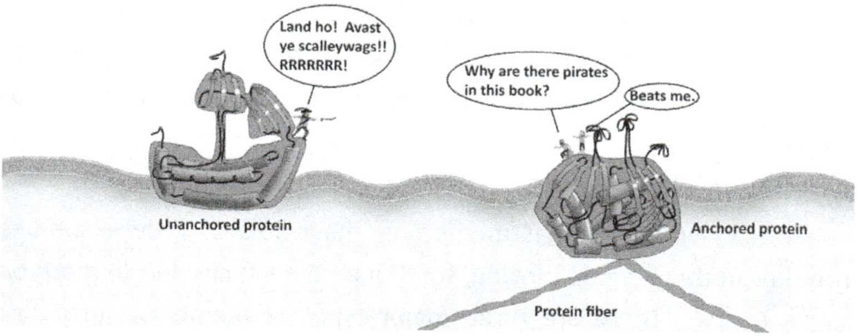

Many membrane proteins have carbohydrate chains attached to them, like the palm trees on the island. The proteins that have these carbohydrates are called **glycoproteins**. These are added in the process called glycosylation, which we referred to briefly in the nucleic acids section of the previous topic text. There are a number of purposes for glycoproteins. First, their carbohydrates have a lot of hydrogen bonding capacity so that they tend to gather and hold a lot of water around them. That makes cell membranes rather mucousy on the outside. Second, glycoproteins serve the purpose of cell recognition, helping other biological molecules know what kind of cell it is, (liver cell vs. heart cell vs. brain cell, etc.).

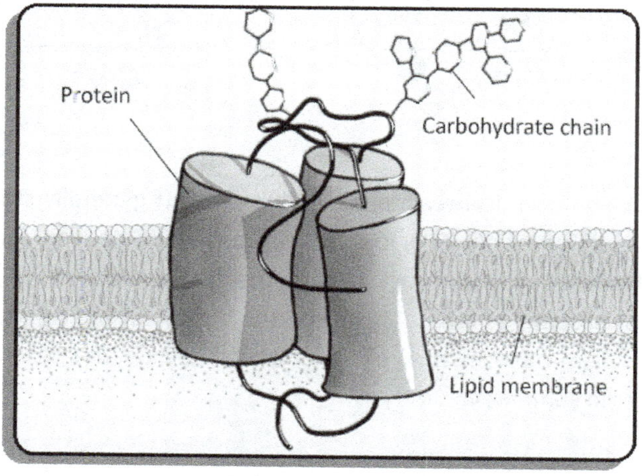

Finally, the lipid bilayer also contains numerous cholesterols (steroid molecules). The more cholesterol that is contained in the membrane, the less stiff the membrane will tend to be. Cholesterols also make the membrane less permeable to little molecules, such as water.

Membrane Proteins

So we know that membranes have lots of little island-like proteins in them, some moving, some not. Let's focus a little more on these things. There are three major types of membrane proteins: receptor proteins, recognition proteins, and transport proteins. **Receptor proteins** bind to very specific molecules that float around in the extracellular fluid, and when these molecules attach, receptor proteins set off a series of chemical reactions within the cell.

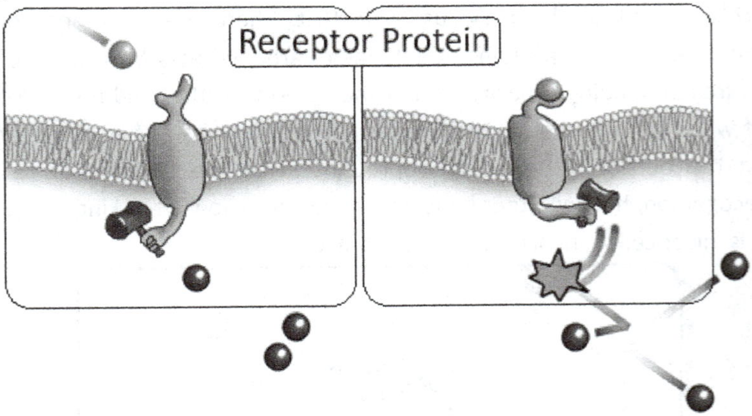

The chemical reactions that they set off generally start with the production of cAMP or cGMP, two nucleotide-derived chemicals we talked about in the last unit of the first topic text. Think of receptor proteins as little switches on the outside of the cell that only certain chemicals can turn on, and when they do the cell changes behavior in some way.

Recognition proteins serve to mark cells, like tags that say what kind of organ they belong to[5] (i.e. lung, intestine, liver, etc.). For multicellular organisms with immune systems, immune cells can cruise around and check for the recognition proteins on a cell to see if it belongs there (friend or foe). In this sense, recognition proteins are like security badges. Furthermore, recognition proteins from neighboring cells can bind to each other and inhibit the cells from growing or replicating before their local neighborhood[6]. This is known as **contact inhibition**. Failure of contact inhibition to stop a cell from reproducing is believed to be one possible cause for cancer.[7] As mentioned above, most recognition proteins are glycoproteins, meaning that they have polymers of carbohydrates sticking out into the extracellular fluid.

Transport proteins serve to move ions and polar molecules across the membrane, which have a hard time passing through the lipid bilayer due to its nonpolar nature. As such, transport

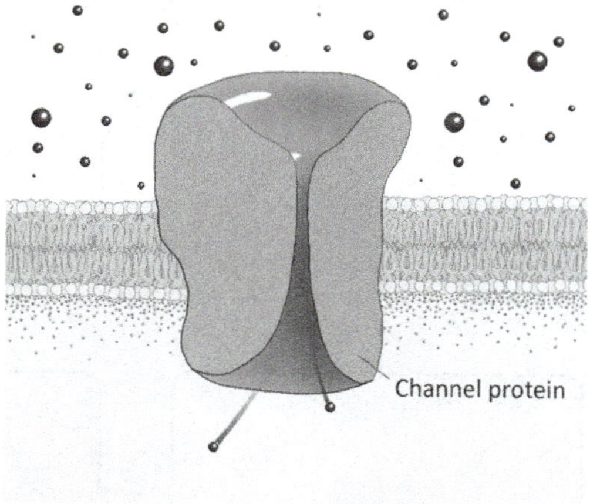

---

5 . . . or is it: "to which they belong?" Bugger me, the English language is so fun.

6 Sadly, *my* neighbors don't have such a preventative mechanism.

7 A cancer is a group of cells that grow and multiply inside the body without regulation, causing problems to the surrounding organs because they take up greater and greater space. After a certain point, some of these cancer cells detach from the original growth and move to other areas of the body. This is called **metastasis**, and is generally the point at which cancer becomes much more difficult to treat medically.

proteins act as specific gates through which some chemicals can pass. Transport proteins are very specific, however, and there are a number of different types. The first type is the **channel protein**. Channel proteins are transport proteins that are basically just hollow tubes through the membrane. The size of the opening usually matches the size of very simple ions, such as K^+ or Na^+, and so these proteins allow only one type of ion to enter or leave the cell.

A second type of transport protein is a **carrier protein**. Carrier proteins bind to specific molecules, and when they do, they change shape and eject the chemical to the other side of the membrane.

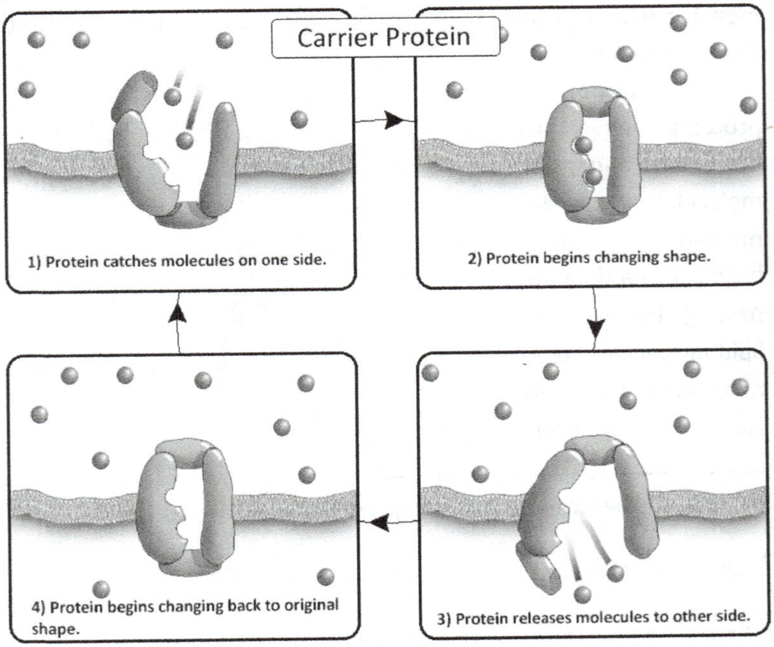

Carrier proteins usually require energy to complete their cycle, and they are often used to force molecules to move backwards up their concentration gradient. The best example of such a carrier protein is the **sodium-potassium pump**. This particular protein grabs

3 Na^+ and throws them out of the cell, and then pulls 2 K^+ ions into the cell from outside. For many multicellular animals, the sodium concentration is higher outside the cell, so sodium would rather be inside the cell. Likewise, the potassium is higher inside the cell, so it would prefer to leave. The sodium-potassium pump thus works to force each type of ion to go in a direction it doesn't want to go. In other words, the Na^+-K^+ pump helps to maintain the sodium-potassium gradients, which would otherwise diminish (as gradients always try to do). Without the Na^+-K^+ gradients, the osmotic strength from *other* chemicals would cause water to rush into the cell. This kind of fate can be observed in cells that have malfunctioning or poisoned[8] Na^+-K^+ pumps; they swell and pop. In addition, cells use the sodium gradient to help pull in sugar, which otherwise would not want to enter the cell. Further, some cells, like nerve and heart cells, use the sodium-potassium gradients as part of their electrical conduction capabilities. That is a topic that is typically studied in the second semester of biology and beyond, so we'll make a graceful exit from that discussion.

One final note on the topic of membranes should be relayed. Cellular membranes aren't only found at the perimeter of the cell. Membranes also surround many of the smaller pieces that are found inside the cell, and these latter membranes typically have the same overall structure and functions. We'll be seeing more about that in later units. More immediately, we'll be taking a closer look at the concept of membrane transportation. It's a complex topic involving more than just transport proteins, so we'll be devoting the next unit to it.

---

[8] Ouabain, an African arrow poison, will bind to the Na^+-K^+ pump and cause it to stop working. This will make the cell expand and burst.

Summary

In this section, we discussed the structure and function of the cell membrane.

- Cell membranes serve three major roles:

  1) They keep the cell components inside, and keep the external world out of the cell.

  2) They selectively exchange materials in and out of the cell.

  3) They communicate with the outside world, notifying the cell if conditions have changed.
- Cell membranes are made of two layers of phospholipids. Molecules of each layer line up so that their nonpolar tails are resting against each other to minimize their exposure to water. The two layers touch each other at the tips of their tails.
- Within the phospholipid membrane are proteins. Some are anchored to a fibrous protein network just inside the cell. These proteins do not move. Other proteins are not anchored and float around freely in the membrane. This is the basis for the fluid mosaic model.
- The phospholipid membranes are also called lipid bilayers. They are very nonpolar. Ions and most polar molecules have a hard time passing through the membranes.
- There are three categories of proteins found in the plasma membrane: receptor proteins, recognition proteins and transport proteins.
- Receptor proteins bind to molecules from the outside and start a chain reaction inside the cell. The chain reaction alerts the cell about the conditions outside the cell.
- Recognition proteins are used to identify the type of cell that it is. Cellular recognition usually leads to contact inhibition, which stops cells from overgrowing their neighbors to

become cancerous. Recognition proteins are also used by immune cells to distinguish friend from foe at the cellular level.

- Transport proteins are used to move chemicals from one side of the membrane to the other. There are two broad categories for transport proteins: channel proteins and carrier proteins. Channel proteins are like a tube that allows specific ions/molecules to pass. Carrier proteins bind to specific ions/molecules, change shape and then release those entities across the membrane.
- The sodium-potassium pump is an important carrier protein. Without it, the sodium-potassium gradients cannot be maintained, and cells pop.
- Membranes can also be found surrounding certain smaller pieces inside the cell. These membranes look and behave in many of the same ways as the membrane surrounding the entire cell.

Unit 3

The Cell Membrane, Part 2: Transportation across the Membrane

In this section we cover:

1) Transportation across the membrane: Passive and Active
2) Passive transportation – simple vs. facilitated
3) Active transportation – protein active transportation, endocytosis (phagocytosis, pinocytosis, receptor-mediated endocytosis), and exocytosis

As we saw in the last unit, the cell membrane is a complex barrier made of a phospholipid bilayer and embedded proteins. Most molecules have a hard time getting through this barrier, so the cell has to have a number of ways to help the chemicals. We talked briefly about transport proteins. In this unit, we expand on the topic of how things get across the membrane, as the full topic is much more complex than simple transport proteins.

There are two main types of membrane transport: **passive transport** and **active transport**. Let's look at them one at a time.

Passive Transport

Passive transport occurs when molecules simply diffuse down their gradients across the membrane. There are two types. The first type is **simple diffusion**. This happens when molecules diffuse across the membrane with absolutely no assistance from the cell; the cell does not need to do anything to get it to happen. We touched upon an example when we talked about osmosis. Water simply diffuses passively across the cell membrane. Other examples include many gases and lipid soluble compounds, (such as cholesterol-based hormones). Gases are usually small, nonpolar molecules, and lipids are nonpolar, so that neither chemical type has difficulty crossing the lipid bilayer. Simply put: simple diffusion is for molecules that do not have trouble getting across the membrane on their own. They diffuse easily, and get into the cell faster than any other molecule, no matter the means.

The second type of passive transport is **facilitated diffusion**. This happens when molecules want to diffuse across a membrane, but need special help in order to do so. This is often the case for most polar molecules and ions flowing through a channel protein. Imagine, for example, a lot of square polar molecules on one side of a membrane and very little on the other side, as depicted in the drawing below. Their concentration gradient is across the membrane, but they can't cross it because the membrane is nonpolar. Now imagine that a door has just been installed in the membrane with a sign that says "squares only." Any square that is floating around can go through the door (i.e. channel protein) and down its gradient, so we say its diffusion is "facilitated."

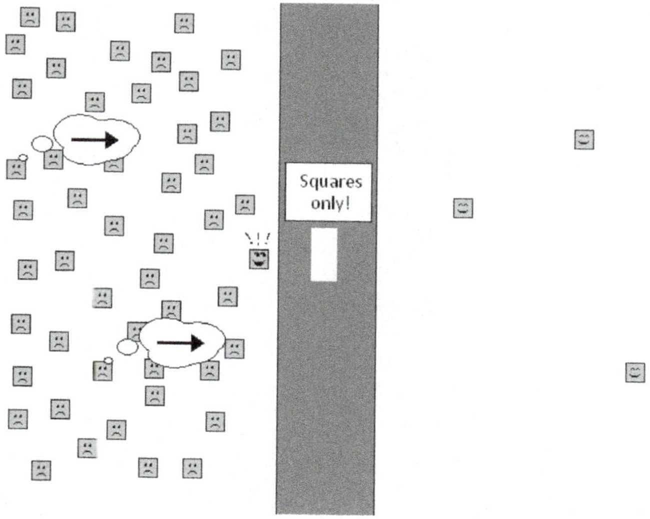

Another form of facilitated diffusion occurs when a molecule diffuses into a cell *against* its concentration gradient because it is paired with a different molecule that is diffusing *down* its concentration gradient. That sounds a bit complicated, so let's take an example. Suppose we reconsider the situation of the squares outside the cell that REALLY want to get in. Now suppose there are triangles inside the cell that want to leave. Let's say the cell doesn't want the triangles to leave. In fact, it wants *more* triangles. This would be the case if the triangle represented a molecule that the cell needs for energy or as a building component. To get more triangles, the cell alters the door to say: "Squares only, but no entry without a triangle." In this way, the cell uses the tendency of one molecule to float in (the square), to force another molecule (the triangle) to come in against its will, so to speak. The diffusion of the second molecule, here the triangle, has been "facilitated" by pairing it with a square.

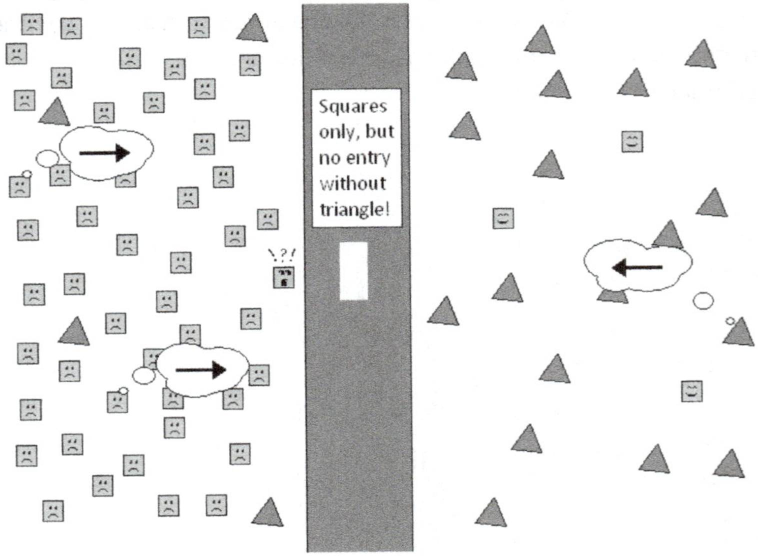

Naturally, cells don't actually have doors with signs on them, so in reality this kind of mechanism is accomplished with a transport protein that does not change shape unless both molecules are bound to it.

For instance, consider the example of Na^+ and glucose. Sodium is like the squares: it wants in. Glucose is like the triangles: it wants out, but the cell doesn't want it to leave. To get more glucose to come in, the cell forces sodium to bring glucose. The cell accomplishes this trick by having a Na-glucose transport protein that binds both chemicals on the outside of the cell. Once both chemicals bind, the protein changes shapes and releases them to the inside.

Active Transport

Passive transport uses the energy locked in chemical gradients. Active transport, on the other hand, relies on the cell's energy to move chemicals across the membrane. There are three broad types: **protein active transport**, **endocytosis**, and **exocytosis**.

We saw an example of protein active transport when we looked at the sodium-potassium pump in the previous unit. The protein goes through a cycle in which it 1) takes up the chemical(s) in question, 2) changes shape, 3) deposits the chemical on the opposite side of the membrane and then 4) changes back into its original shape for another cycle. This cycle takes chemicals and moves them backwards up their gradients. Since this goes against the gradient, it requires energy to do so. To power the cycle, cells use high energy molecules (usually ATP, see Unit 11 of Topic Text A). These high energy molecules bind temporarily to the protein pump, give it the energy it needs, and then break off exhausted and spent like a dead battery. In the case of ATP, it breaks into ADP and phosphate during the process of charging the protein.

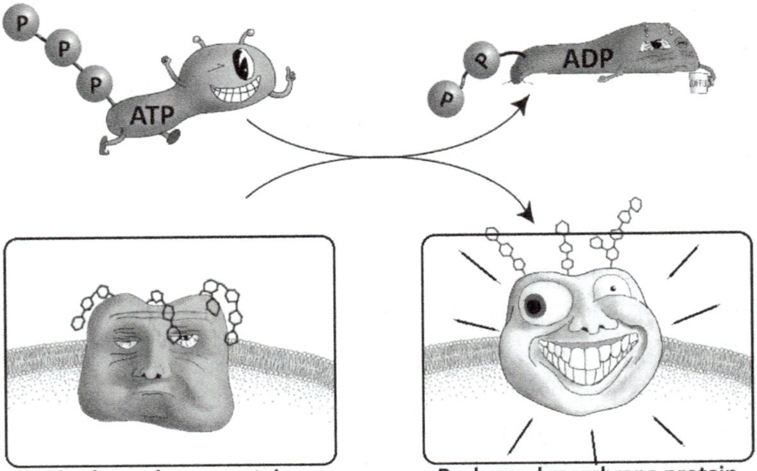

Tired membrane protein Recharged membrane protein

As this image shows, ATP gives its energy to the protein, and the protein is recharged to do its job. . . and for some reason looks like me after four caffè lattes. Anyway, if the above protein was a charged sodium potassium pump, it would eject 3 Na^+ from the cell and pull 2 K^+ into the cell before becoming tired again.

Whereas protein active transport involves small molecules pumped a few at a time across a membrane by a simple protein, endocytosis and exocytosis involve the mass transport of BIG things into and out of the cell, respectively. By "BIG things," I refer to objects such as another cell or a piece of a cell that is too large for protein transport to handle, and certainly too large for any other means of diffusion. There are a few different types of endocytosis, depending on what is being taken in and the mode of delivery. Those types are **phagocytosis**, **pinocytosis** and **receptor-mediated endocytosis**. Phagocytosis is akin to the cell "eating" a large thing floating by in the extracellular fluid. The cell senses the object through various protein interactions at its surface (the details being beyond the scope of this book), and slowly wraps its membrane around the object, much like a mouth eating a piece of food. When the membrane wraps around the object entirely, it joins together, creating a membrane pouch (or **vesicle**) around the object, which is now inside the cell. Once in this state, the cell employs a variety of means to break down the object. The pieces that come off of the object leave the pouch by simple diffusion, facilitated diffusion, or protein activated transport.

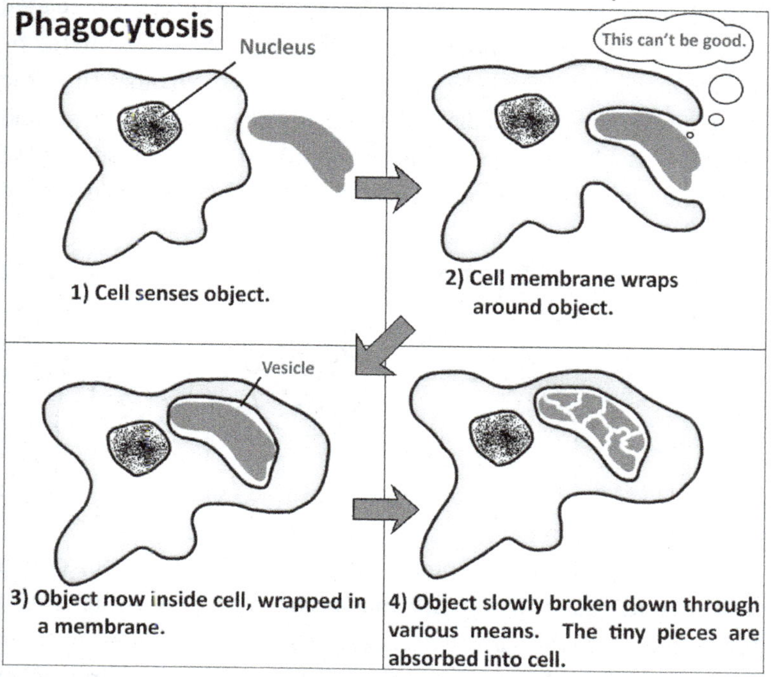

Pinocytosis is a similar process, but instead of the membrane wrapping around an object, it wraps around a big section of extracellular fluid, like biting water. As with phagocytosis, this creates a membrane-bound vesicle inside the cell. Inside the vesicle are extracellular fluids and whatever chemicals were in it. The materials in the pouch then diffuse through the membrane into the cell, (like it does in phagocytosis).

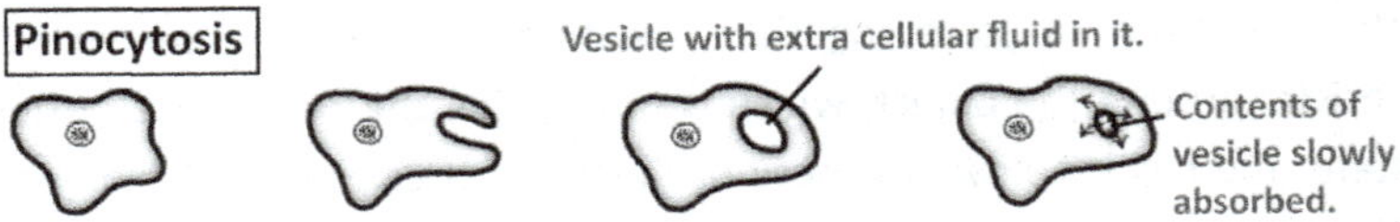

Pinocytosis is considered non-specific. The cell is simply gulping down a volume of extracellular fluid. Receptor-mediated endocytosis is a similar process, but unlike pinocytosis it is initiated by specific molecules binding to various receptor proteins that sit in the cell membrane. The binding of these molecules causes a protein network (usually proteins called **clathrin**) to pinch a section of membrane inward. This pulls a vesicle full of the extracellular fluid into the cell.

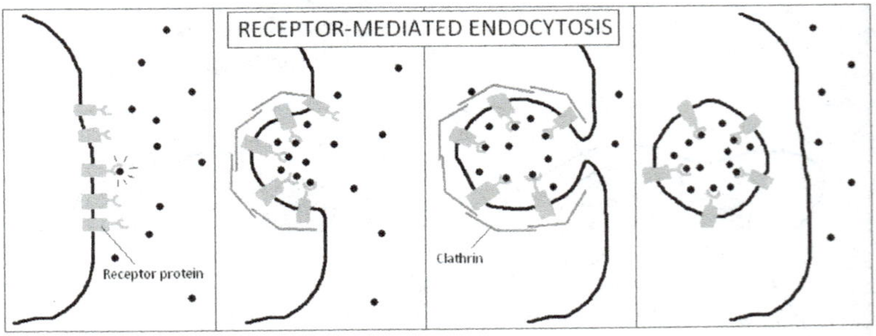

Exocytosis is the opposite of endocytosis. In exocytosis, a vesicle combines with the cell membrane from the inside, like a bubble coming to the surface of water. When such a bubble pops, the air inside the bubble is released to the atmosphere. Similarly, when the vesicle combines with the membrane and "pops," the material inside the vesicle is released to the outside of the cell.

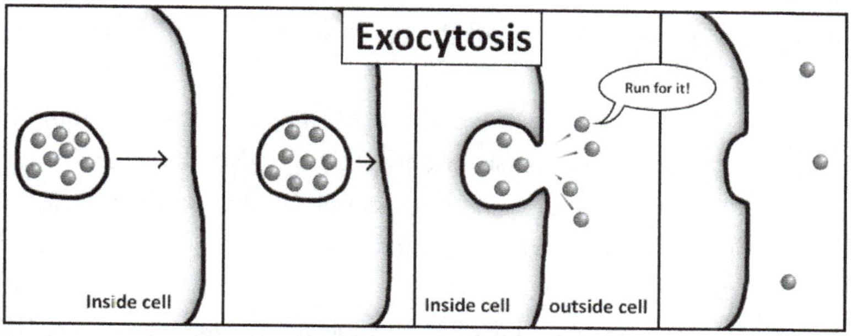

With that, we bring to a close our discussion of the basic methods of transporting materials across the cell membrane. In the next section, we'll conclude our discussion of the cell membrane by taking a look at plant cell walls, bacterial walls and a few other additional structures found near the membranes.

Summary

In this section, we talked about the basics of transportation across cell membranes.

- There are two broad categories for membrane transportation: passive transport and active transport.
- Passive transport uses the energy in chemical gradients to help chemicals cross membranes. There are two types of passive transport.
    - Simple diffusion – molecules that can cross the membrane on their own, like water and gases. This is the fastest means that chemicals enter or leave cells. The most important example is osmosis, (the movement of water).
    - Facilitated diffusion – protein channels help chemicals diffuse down their gradients. Sometimes one molecule diffuses down its gradient, dragging another molecule up its gradient, like Na^+ and glucose into the cell.
- Active transportation requires the cell's energy to get activated. There are several types.
    - Protein active transport – uses proteins to pump molecules up their gradients. These pumps require high energy molecules like ATP to keep charged.
    - Endocytosis – large scale movement of materials across the membrane via vesicles that are made when the membrane surrounds the materials.
    - Phagocytosis – the cell folds its membrane around a large object, (such as another cell), and pulls it in.
    - Pinocytosis – the cell gulps a volume of extracellular fluid by wrapping the membrane around it, and then pulls it in.

- Receptor-mediated endocytosis – specific chemicals bind to receptors in the cell's membrane, activating a process in which the membrane invaginates inward to make a vesicle. Clathrin is a typical protein that helps to make the vesicle form.
- Exocytosis – this is the opposite of endocytosis. A vesicle inside the cell combines with the membrane and then pops to release its contents outside the cell.

Unit 4

The Cell Membrane, Part 3: The Cell Wall and Cell Coat

In this section we cover:

1) Cell Wall, basic purpose
2) Types of cell walls: plant, fungi, bacterial
3) Gram positive and gram negative bacteria
4) The glycocalyx structure and function

Cell Walls

In addition to the cell membrane, some cells have a tough **cell wall**. These typically occur in plants, fungi and bacteria. The cell wall creates a strong barrier that protects the cell against a variety of physical forces, including the threat of popping when too much water flows in by osmosis. When this happens, the water coming in causes the cell to swell up and push against the inside of the cell wall, building what is called **turgor pressure**. The turgor pressure stops more water from coming in, allowing organisms that use cell walls to have much more resistance to environments that are hypotonic. Further, when the turgor pressure builds up, the entire cell (with cell wall) becomes stiff, giving the organism greater structural strength. This is important in plants.

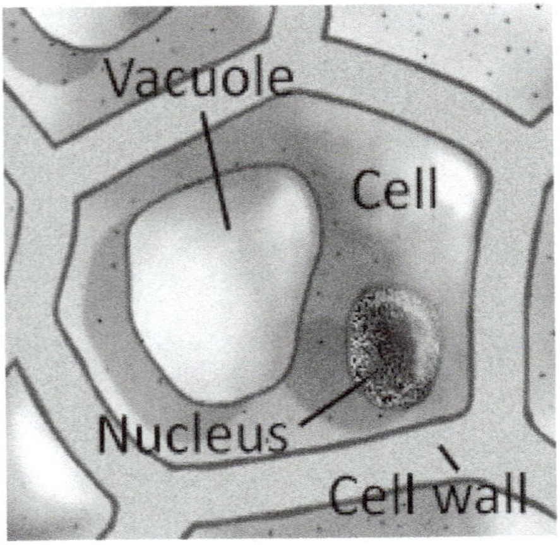

The cell wall provides the basis for structural support for plants and many macroscopic fungi. By surrounding the cells in a tough casing, the entire organism can withstand the effects of gravity. In this sense, the cell walls of plant cells act like a rigid skeleton that keeps plants and fungi from crumbling into a massive pile of goo.

In this last image, I've drawn a little compartment inside the cell called a **vacuole**. Vacuoles are little sacs inside plants cells that act (in part) as waste dumps for the cell. Instead of having a sewage system that takes cell garbage away, the plant cell just puts its garbage in the vacuole. On that note, it's a shame you don't know my roommate, because I had a great analogy for this, but then again if you've ever lived in the vicinity of a college dorm, you probably already know what I mean. Anyway, there is a secondary purpose to the vacuole: as it swells up, it helps to establish the turgor pressure that contributes to the overall rigid structure of plants.

Plant Cell Walls

The structure of cell walls depends on the type of organism. Plant cell walls are made of primarily cellulose, the long-stranded carbohydrate we saw when we discussed carbohydrates in the previous topic text. The sugar chains in cellulose are lined up next to each other, making a sheet. Then another sheet is laid on top of it in such a way that the strands face a different angle (usually 60 to 90 degrees from the previous layer), so that the result is a netlike structure. It's similar to thatched baskets, except that the carbohydrates are not interwoven.

Holes and gaps are frequently left in these sheets, so that water, air and other small molecules can freely pass through. Plant cell walls also have other carbohydrate compounds in them, such as **pectin** and **hemicelluloses** (show below). These occur in lesser amounts and are used to hold the cellulose strands together.

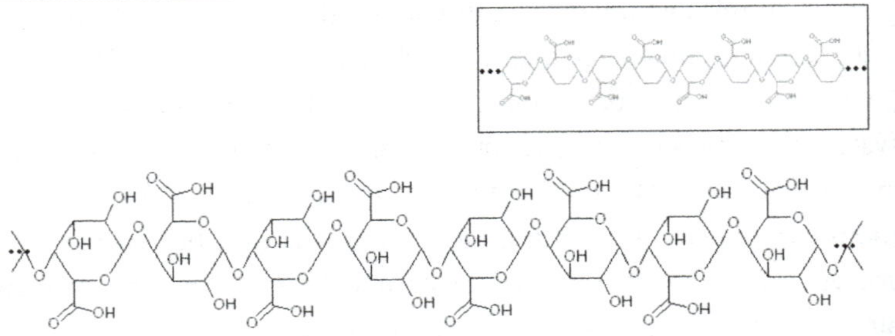

Pectin

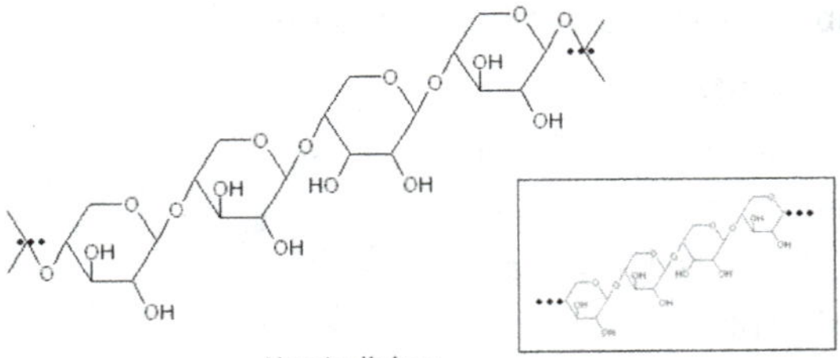

Hemicellulose

When plant cells make their walls, they first lay down a **primary cell wall**. This particular wall is somewhat elastic and capable of expanding as the cell grows. When the cell stops growing, it produces a **secondary wall** that is thicker and more rigid. This secondary wall is deposited *inside* the primary wall, and unlike the primary wall, it contains a significant amount of lignin (a complex molecule that does not readily fall into the four primary macromolecule types we discussed). Where two plant cells make contact, there is a connecting space called the **middle lamella** that has a lot of pectin in it. When the middle lamella breaks down, such as when fruit rots, the cells break free of each other, making the entire collection of cells mushy.

At the intersection of multiple plant cells are air spaces that form long tubes. These help deliver various atmospheric gases throughout the plant.

Plant cell and wall layers

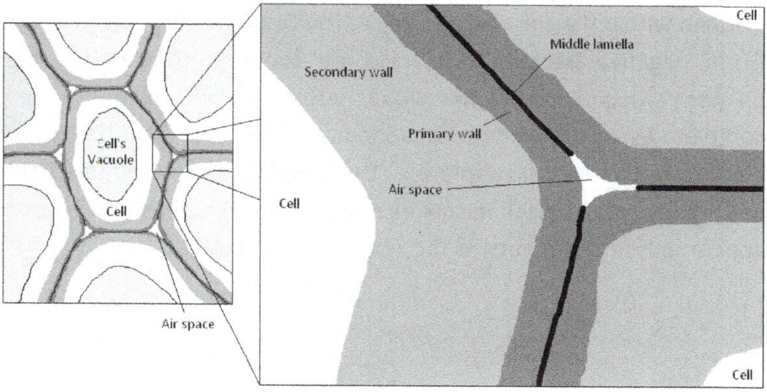

Fungal Cell Walls

Fungal cell walls are made from **chitin**, the polymer created from monomers of glucosamine we in Unit 8 of Topic Text A. Chitin is the same material that is used in the shells of insects and crustaceans (crabs, shrimp, lobsters, pop singers, etc).

Bacterial Cell Walls

Bacterial cell walls are made from **peptidoglycans**, which are basically long chains of polysaccharides cross-linked by short proteins. **Penicillin** kills bacteria by inhibiting the production of the bacterial cell wall, making the bacteria susceptible to osmotic changes. The result is that they pop and die. Broadly speaking, there are two types of bacterial cell walls: **Gram positive** and **Gram negative**, named after a staining technique used to see bacteria under microscopes.

Gram positive bacteria have thick cell walls containing a great deal of teichoic acids in them. Teichoic acids are large polymers of variable structure that commonly have a lot of repeating units of glycerol and phosphate. They attract a lot of metal ions and provide the wall with increased toughness. Gram negative bacteria have thin cell walls with no teichoic acid. They also have a second lipid bilayer membrane surrounding them. The Gram staining process is rather complex, but it works essentially like this: A blue pigment gets trapped within the cell wall and cell of the Gram positive bacteria. The Gram negative bacteria cell walls are too thin to trap the blue pigment, which then washes away. After the blue pigment step is completed, a red pigment is added. This makes the blue Gram positive bacteria turn purple and the colorless Gram negatives turn red. A few other mechanisms also occur to cause reds and blues to appear, but they are outside the scope of this book.[9]

Cell Coats

The cells of animals do not have cell walls. Instead, they have a layer of short carbohydrates that are covalently bound to the lipid bilayer[10]. This forms a hairy coat-like covering around the cell that is called the **glycocalyx**. As we saw when we talked about

---

[9] For example, some Gram positive cells are too fragile for this process and break, causing them to lose their blue pigment and appear red at the end. Altogether, the process can get complicated and is best left for courses on microbiology.

[10] Until now these short carbohydrates have only been drawn attached to random membrane proteins (i.e. glycoproteins).

carbohydrates, sugars are very polar, so the glycocalyx attracts a great deal of water to the surface of the cell, making it very slimy. The primary purpose of the glycocalyx is to provide recognition[11]. Each organ type, for example, has a different mix of oligosaccharides in its coat, so heart cells, and kidney cells, and intestinal cells will each have different glycocalyx coats. Secondarily, the glycocalyx serves as a form of contact inhibition. This means that nearby cells sense the glycocalyx of another cell and stop growing if there is not enough space. If they lacked this preventative measure, the cells would continue to grow and divide out of control, forming a cancer.

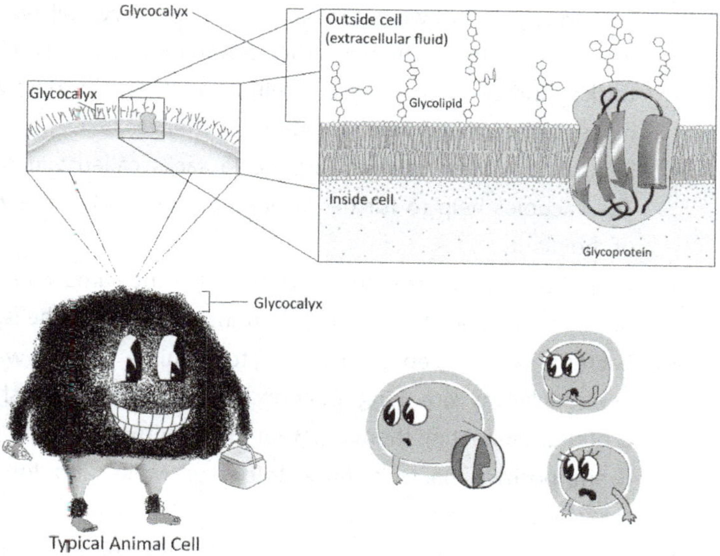

This brings us one step closer to closing our discussion about the outside of cells. We've got one more topic to cover about cell membranes, then we'll transition into talking about the insides of cells and how they work.

---

11 . . . just like glycoproteins do!

Summary

In this section, we discussed cell walls and the glycocalyx.

- Cell walls are tough outer coverings that protect cells from physical stresses as well as osmotic popping from hypotonic environments.
- Water that flows into cells will make them swell up against their cell wall. This creates turgor pressure that prevents further swelling and strengthens the overall structure.
- Plants have two types of cell walls: the primary cell wall (the outermost) laid down first, and the secondary cell wall (the innermost) laid down afterwards. The primary cell wall is more elastic and is thinner. The secondary cell wall is much thicker and contains a lot of lignin in it. It is very rigid and helps to give plants their rigid stems and leaves.
- Plants store their cellular garbage in sacks called vacuoles. The vacuoles help to further strengthen the rigid framework of the plant.
- Fungi have cell walls made of chitin. It is the same material used by bugs and crustaceans when they make their shells.
- Bacteria use peptidoglycans to build their walls. Peptidoglycans are long polysaccharide chains with short proteins connecting them together into a network.
- Gram positive bacteria have thick cell walls and lots of teichoic acid in them.
- Gram negative bacteria have thin cell walls and no teichoic acid. They also have a second cell membrane on the outside of the wall.
- Gram negative bacteria are named because the gram stain does not stick well to them. This is mainly because their small cell wall does not trap the dyes very well.

- Many antibiotics, like penicillin, work by disrupting the cell wall, causing the bacteria to burst.
- The glycocalyx is the furry polysaccharide coat found on many types of animal cells. It gives the cell a slimy covering, provides it with a distinctive coat that tells other cells what kind of cell it is, and helps to give the cell contact inhibition so that it does not overgrow its neighbor cells.

Unit 5

The Cell Membrane, Part 4: Cell Membrane Structures

In this section we cover:

1) Spot Desmosomes
2) Belt Desmosomes
3) Tight Junctions
4) Gap Junctions
5) Plasmodesmata

In the previous unit, we saw that neighboring cells can detect each other by sensing their oligosaccharide glycocalyx coat on the outside of cells. There are a few other structures that cells use to communicate with neighboring cells. These include the **spot desmosome**, the **belt desmosome**, the **tight junction**, the **gap junction** and (in plants) the **plasmodesmata**. These structures are shown in the large image on the next page. The image is meant to be a generic cell; not all cells have all these structures.

Spot Desmosome

Spot desmosomes are coin-like structures that sit on the inside of the plasma membrane opposite similar structures of neighboring cells. They have proteinaceous threads (made of a membrane protein called **cadherin**) that attach them to the opposing desmosome, running through the plasma membranes of the cell and its neighbor. On the cytoplasmic side of these disks are other protein threads made of **keratin**. The keratin is part of the internal network of fibrous threads called the **cytoskeleton**, which we will discuss later. Spot desmosomes are commonly described as little rivets that help hold neighboring cells together.

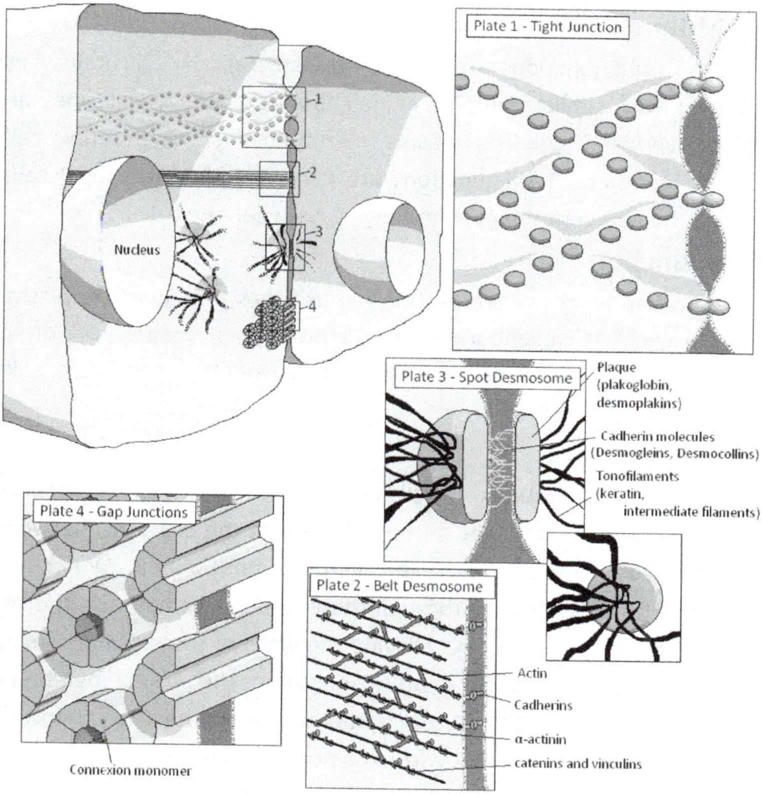

Belt Desmosome

Belt desmosomes are lines of proteins that sit in the plasma membrane and form a ring around the circumference of the cell. These proteins are connected with a network of **actin** proteins on the inside of the cell. Actin, as we will see, is a common protein involved in contraction (such as in muscles). The exact roles of belt desmosomes are still debated, but they are believed to serve as an internal support for the cell, like a regular array of hooks from which you could string ropes across a room, say.

Tight Junction

Tight junctions are lines of transmembrane[12] proteins that connect with similar proteins in neighboring cells. These lines are interwoven and pull the plasma membranes of neighboring cells tightly together. Tight junctions are very typical of intestinal cells, where they prevent things from flowing between the cells.

Gap Junction

Gap junctions are proteinaceous tubes that span the plasma membranes of neighboring cells, forming a means of direct communication between the cytoplasms of the two cells, like underwater tunnels.

Plasmodesmata

In plants, tough cell walls provide the majority of cell support. Communication is achieved through breaks in the cell walls between adjacent cells; these breaks are called plasmodesmata. Water and nutrients can flow directly through these breaks. Sometimes, the cell membranes of the adjacent cells line the walls of the plasmodesmata and meet each other at the center. Sometimes the plasma membrane melds directly with the plasma membrane of the next cell, allowing the cytoplasm to flow directly from one cell to another.

---

[12] Transmembrane = a membrane that spans across the membrane. This word describes most membrane proteins, actually.

Plant Cell and Wall layers

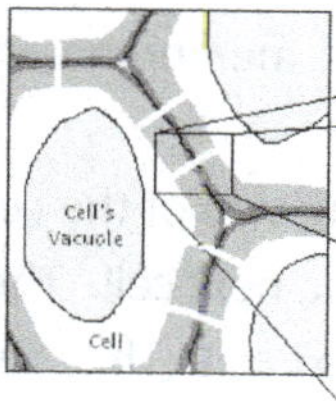

Close-up of Plasmodesmata

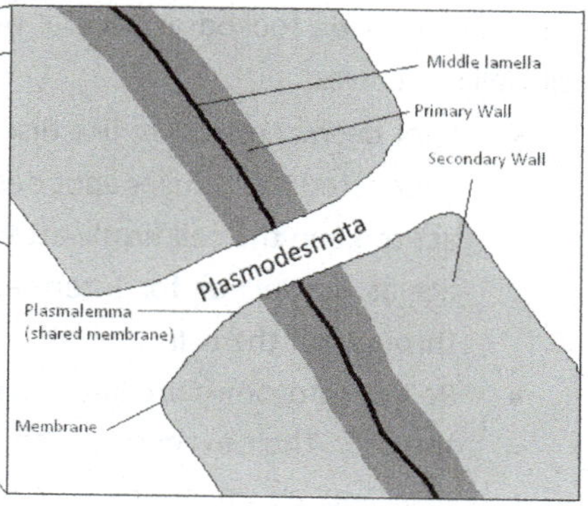

Summary

In this section, we looked at several of the basic cell membrane and cell wall structures.

- Spot desmosomes are like discs welded to the side of the cell, connected to a similar spot desmosome in the next cell. They act to keep the cell firmly attached to the next cell. They also act as an anchor for intermediate protein threads that run throughout the cell.
- Belt desmosomes are lines of actin proteins that circle around the cell. Their exact role is still debated, but they are thought to act as an anchoring place for other protein threads that run into the cell.
- Tight junctions are interwoven lines of proteins that tightly bind the membrane of one cell with the membrane of another. They prevent material from seeping between cells and are very important in hollow organs like the intestines.
- Gap junctions are little protein tunnels that directly connect the cytoplasm of one cell with another. They allow materials to transfer rapidly between cells.
- Plasmodesmata are openings in the cell wall of plants that allow the plant cells to communicate with each other. The plasma membranes that run through these openings can either be fused so that the cytoplasm flows directly, or they may lay against each other forming a double-membrane barrier midway across the plasmodesmata.

Unit 6

Cell Anatomy

In this section we cover:

Organelles:

1) Nucleus
2) Nucleolus
3) Chromosomes
4) Ribosomes
5) Smooth Endoplasmic Reticulum
6) Rough Endoplasmic Reticulum
7) Golgi Apparatus
8) Lysosomes
9) Peroxisomes
10) Mitochondria
11) Chloroplasts
12) Microfilaments
13) Intermediate filaments
14) Microtubules
15) Centrioles
16) Cilia
17) Flagella

At this point, we've talked about the cell membrane at great length. We discussed how it transports molecules and materials into and out of the cell and conversed over the kinds of structural things you might find at the cell surface.

We will now discuss the many things you can "see" inside the cell. First, "see" is a bit of a tricky word to use. The structures we are going to talk about have no color at all. If you plop a cell under a microscope in an attempt to marvel at its many innards, you'd be lucky enough just to see a vague circle, or oval, or hotdog shaped thing that represented the cell as a whole. The only way to actually see the stuff inside the cell is to stain the cell with some kind of molecule that has color. The stain you use depends on the structure you want to see. We use a whole slew of different colored molecules, each having some tendency to attach to a particular type of structure inside the cell. Over the years, we've created newer ways to see inside the cell, including fluorescence and electron microscopy. As these more advanced imaging techniques emerged, we've been able to look at these structures in new ways, increasing our impressive understanding of the cellular workings of life.

When you look at a cell under a microscope, you won't always see all the structures that we will be discussing in this unit. There are many reasons for this; the first one we already mentioned. If you don't include the proper dye, you won't see a particular structure because it has no inherent color. Also, cells are 3D structures. Microscopes only focus at one depth at a time. So, if you are focused on one particular depth, say, just below the cell's surface, you will not see the structures that are positioned much farther down. Finally, there are a wondrous variety of different cells out there in nature, and not all cells have all the structures in their arsenal of pieces-parts. Many of the structures that we discuss here are fairly universal though. So, without further delay, let's begin our discussion of the anatomy of the cell.

The Organelles

Just as you have organs that make up your body (i.e. the heart, the intestines, the kidneys, and for most of us the brain), the cell has individual parts as well. These we call the organelles. We talked about two already: the cell membrane and the cell wall. As you know the cell has a membrane around it, but does some of the organelles inside the cell, as well! The most obvious of these, (and typically the first organelle to learn about), is the **nucleus**. The nucleus is the headquarters of the cell where the DNA is stored. Surprisingly, not all cells have a nucleus. The presence or absence of a nucleus allows scientists to divide all organisms into two broad categories. The first category is the **eukaryotes**, which have nuclei. All plants, animals (including humans), fungi and some microscopic organisms are eukaryotes[13]. The second category is the **prokaryotes**, which do *not* have nuclei. Their DNA is mixed in with the rest of their cytoplasm. Prokaryotes are typically believed to be more primitive forms of life, even more so than my roommate. Unlike my roommate, however, nearly all prokaryotes are microscopic organisms. Bacteria are an example of prokaryotes.

On the next page is a list of the major organelles of the cell, as well as their general function. Refer to the image that follows for an idea of what they look like. It also shows the visual differences between eukaryotes and prokaryotes.

---

[13] Technically, many organisms, such as animals, have specialized cells inside them that have had their nuclei removed, such as the case with red blood cells. Premature red blood cells have nuclei, but the nuclei and DNA are eventually kicked out of the cell when the blood cell matures. Even though the blood cell has no nucleus, it is still considered a eukaryotic cell, since it *started* with a nucleus.

Organelle	Major Function(s)	Eukaryotic/Prokaryotic
Cell Membrane	Barrier to retain cytoplasm Cell recognition Material Transportation	Both
Cell Wall	Strong structural support Prevents osmotic bursting	Eukaryotes (plants, fungi, algae), Prokaryotes
Chromosome	A large polymer of DNA used to store genetic information Becomes discreet and observable in eukaryotes when the cell divides	Both
Chloroplast	Membrane bound organelle in plants where light is captured and turned into chemical energy Has its own unique set of genetic material	Eukaryotes (plants and protists)
Flagellum (plural flagella)	Protein whip-like structure that protrudes from the cell and serves as a propeller, pushing the cell through the water	Both
Golgi Apparatus	Membrane bound organelle responsible for the processing of material destined to be secreted from the cell Makes lysosomes	Eukaryotes
Intermediate filaments	Small molecular threads used for structural stability.	Eukaryotes
Lysosome	Involved in the breakdown cell waste, engulfed objects, worn-out proteins, carbohydrates and lipids Involved in the transport of the cell waste to the cell surface for removal	Eukaryotes (mainly animals)

Organelle	Major Function(s)	Eukaryotic/Prokaryotic
Microfilaments	Small molecular threads used for structural stability	Eukaryotes
Microtubules	Small molecular threads used for structural stability, cell material transport, cell movement, and cell division	Eukaryotes
Mitochondrion (plural: mitochondria)	Membrane-bound organelle that serves as the energy harvesting structure of the cell Has its own set of unique genetic material	Eukaryotes
Nucleolus	Circular shape within the nucleus seen when the cell is not dividing Contains RNA, active DNA, and various proteins used to create protein blueprints called messenger RNA, or mRNA	Eukaryotes
Nucleus	Contains chromosomes (the genetic material, DNA) Surrounded by two membranes that separate it from the cytoplasm	Eukaryotes. The presence of a nucleus is the defining feature of Eukaryotes
Plasmid	Small circular DNA molecules found in many bacteria. This is in addition to their large circular chromosome.	Prokaryotes, though they can also be found in chloroplasts and mitochondria
Peroxisome	Vesicle that metabolizes long chain fatty acids, less commonly found amino acids, and some sugar molecules. Uses peroxide in its reactions	Eukaryotes

Organelle	Major Function(s)	Eukaryotic/Prokaryotic
Ribosomes	Large protein/RNA structures within the cytoplasm that are responsible for the manufacture of proteins	Both
Rough endoplasmic reticulum	Same as smooth endoplasmic reticulum, but with ribosomes attached to it Involved in the initial stages of protein secretion (transportation of proteins out of cell)	Eukaryotes
Smooth endoplasmic reticulum	Tubular system that is continuous with the nuclear double membrane Stores and sorts materials for transit throughout the cell or directs materials to be secreted from the cell	Eukaryotes

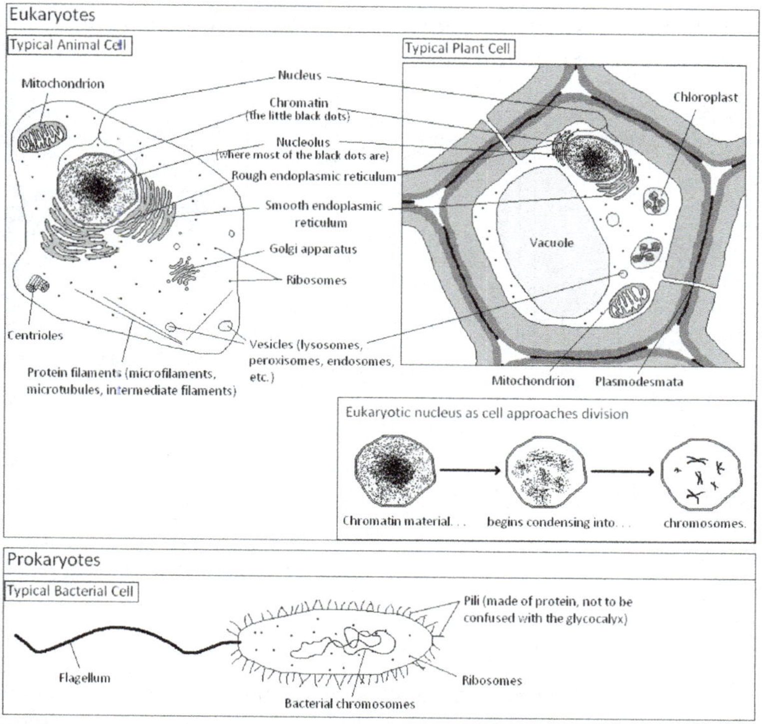

The Nucleus, Chromosomes, and Nucleolus

The headquarters of the cell, the nucleus stores the DNA and is separated from the cytoplasm by a double-membrane called the **nuclear membrane** or **nuclear envelope**. Each polymer of DNA is wound tightly upon lots of protein complexes called **nucleosomes**[14]. The nucleosomes look like yo-yos with the DNA wrapped twice around each one. These protein yo-yos are then attached to each other to create a compact spiraling structure that is tightly wound. This spiral is, in turn, gathered into a densely packed arrangement to form a **chromosome**. A great many proteins are required to complete these

[14] Nucleosome actually refers to the protein complex (made of 5 smaller proteins called histones) as well as the DNA wound around it.

structures, but they are outside the scope of this book. Here is a simplified view of how chromosomes are organized.

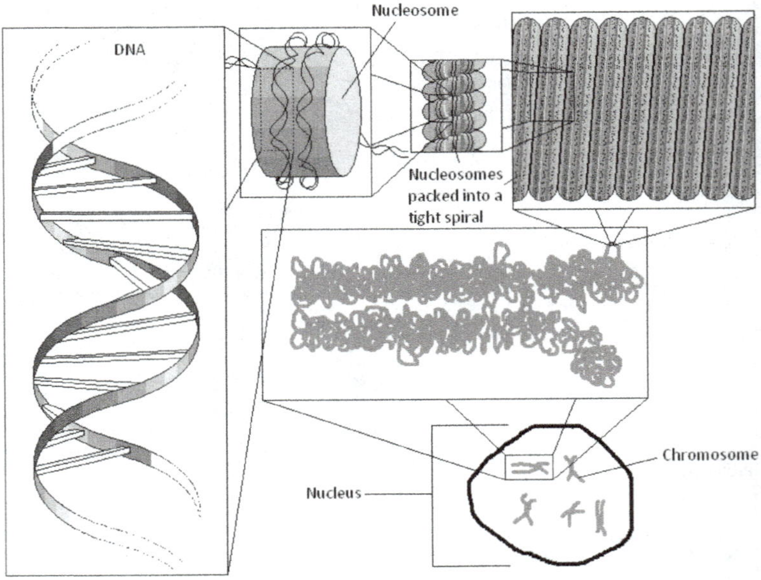

When the cell is dividing, the chromosomes look similar to how they are shown in the bottom right of this latest image. With the proper dye, you can see discreet, individual chromosomes, and if the focus is lucky, you might be able to see them all at once. The total number of chromosomes that a cell has, however, depends on the species. Humans, for example, have 46 total chromosomes[15].

When the cell is *not* dividing, the chromosomes are less tightly packed, and under the microscope they can appear as cloudy material within the nucleus. This cloudy material is called **chromatin**. The darkest part of this material, usually in the center of the nucleus, is called the **nucleolus**. The nucleolus is a collection of active DNA, RNA, and various proteins that are interacting during the preparatory

[15] More specifically, we have 23 pairs of chromosomes. We'll learn more about this later.

stages of protein production, which we will discuss in the next topic text. In short, "active" DNA is DNA that is being used to make the blueprints for protein. All the molecules involved in this process are gathered in one place, making the nucleolus look darker and more crowded than the rest of the nucleus. The blueprints that are made are shipped out of the nucleus; the actual protein production takes place in the cytoplasm.

Cell Nucleus

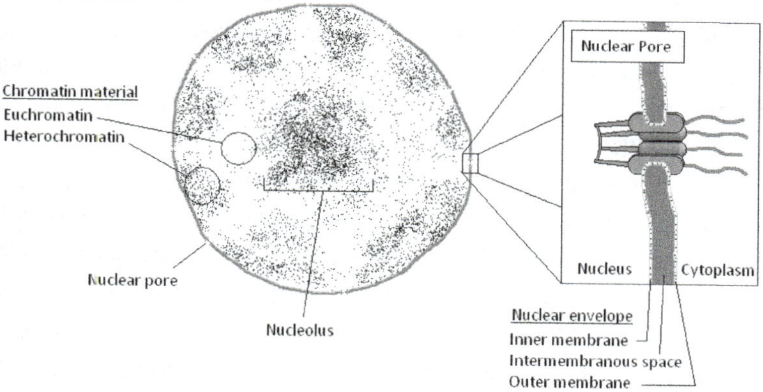

As this latest image shows, the chromatin outside the nucleolus is divided into two categories: **euchromatin** and **heterochromatin**. Euchromatin refers to the areas where the DNA is less dense. It forms little canals between the denser heterochromatin areas. The heterochromatin that is pushed up against the inside of the nuclear membrane is the DNA that is not currently being used.

Also in this image, notice the small openings in the nuclear membrane. Within these openings are protein structures called **nuclear pores**. These protein complexes help various large molecules get into and out of the nucleus.

Unlike eukaryotes, prokaryotic chromosomes are not as tightly wound and are typically circular in structure, as if the ends of the chromosomes had been sown together. Furthermore, prokaryotes typically have additional small circular DNA molecules called **plasmids**. These are basically small packets of DNA that carry information for extra proteins that confer additional abilities to the bacteria. We'll encounter them again much later.

Ribosomes

Ribosomes are structures that float around in the cytoplasm of the cell. They are large compared to most proteins, but still small enough that they are little more than dots inside the cell. Ribosomes are divided into a large subunit and a small subunit. Together they look like little gloves interlocking each other. Each subunit is made from both protein and RNA. This RNA is called **rRNA**, which stands for **ribosomal RNA**.

The purpose of ribosomes is to synthesize protein, a process we will look at more in depth in the next topic text. For the moment, a quick overview is as follows: RNA polymers called **messenger RNA**, or **mRNA**, are made in the nucleolus using information in the DNA. These act as the "protein blueprints." The mRNA finds its way out into the cytoplasm where it encounters a ribosome. It then travels through a channel between the subunits of the ribosome, interacting with various cellular molecules inside the ribosome to make protein. The new protein slowly appears from the other end of a tunnel that pierces the large subunit. Here is an image to help clarify the structure and concept:

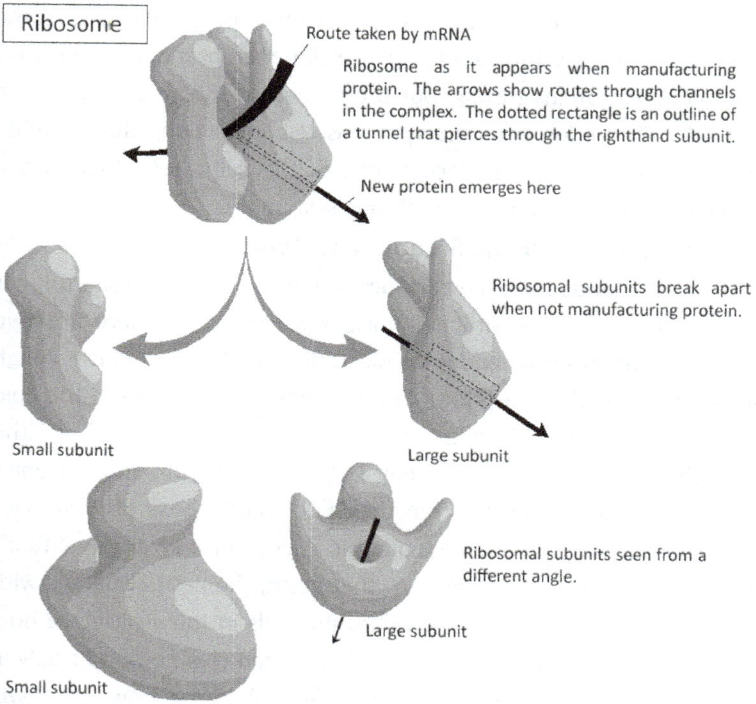

As we will see shortly, ribosomes can be found in two different places. First, they can be found floating around in the cytoplasm. The proteins they make there will be released into the cytoplasm. Second, they can also be found attached to the endoplasmic reticulum (discussed below). The protein they make there get put inside the endoplasmic reticulum and routed to various vesicles. It may then eventually be exported from the cell.

The Endomembrane System

The **endomembrane system** is the name given to the collection of organelles that include the smooth and rough endoplasmic reticulum, Golgi apparatus and the various vesicles that transit about the cell. The nuclear membrane is technically also a member of the endomembrane system, though a discussion of it was more appropriate in relation to the nucleus (above). At any rate, the endomembrane system works to transfer membrane phospholipids and various proteins throughout the numerous membranes found in the cell. The endomembrane system is also involved in the secretion of proteins and other molecules out of the cell. Let's look at the individual parts of the endomembrane system.

Endoplasmic Reticulum, Smooth and Rough

The **endoplasmic reticulum** is a series of membrane tubules that connect directly with the nuclear membrane. There are two varieties, the **smooth endoplasmic reticulum (SER)** and the **rough endoplasmic reticulum (RER)**. The phospholipids that are in the lipid bilayer membranes of the cell are made in the SER. As the phospholipids are produced, they begin to form into little bubbles called **vesicles** that bulge from the SER. Then, these little vesicles depart from the smooth endoplasmic reticulum and are eventually taken to the cell's membrane. At that point, the vesicle merges with the cell membrane, so that the phospholipids in the vesicle are now part of the cell membrane. Since the lipids can roam around freely in the membrane (i.e. the fluid mosaic model), these new lipids will diffuse across the membrane and all around the cell.

The SER also stores calcium, which is important for processes such as flexing muscles, a topic usually covered in the second semester of biology courses.

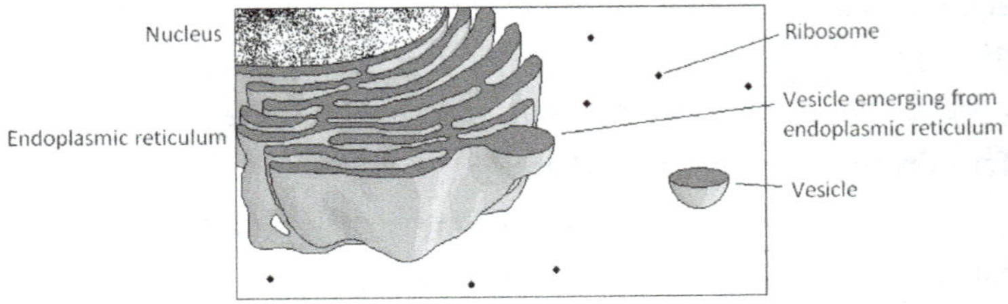

The rough endoplasmic reticulum is like the SER, but with ribosomes attached to its surface. These ribosomes make proteins and thread them into the hollow inside (i.e. the **lumen**) of the endoplasmic reticulum. Alternately, the ribosomes attached to the SER embed the new proteins directly into the membrane of SER.

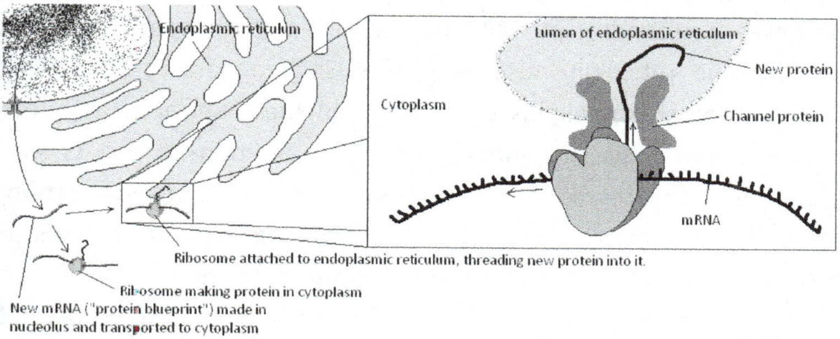

When a vesicle forms from the endoplasmic reticulum, the proteins that are floating in the lumen get trapped inside the vesicle. Likewise, the proteins that were embedded in the membrane of endoplasmic reticulum become part of the membrane of the vesicle. When the vesicle departs for the cell membrane (or other destination), the materials inside it and the proteins in its membrane tag along for the ride. As the vesicle combines with the cell's membrane, the proteins inside the vesicle get released to the outside and the proteins embedded in the vesicle's membrane become part of the cell's membrane proteins.

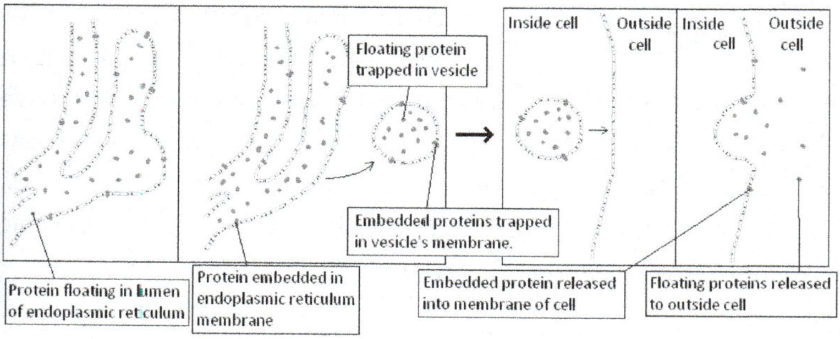

The vesicle's trip is actually slightly more complex though. Before it reaches the membrane, it first goes through the **Golgi Apparatus**, so let's look at that.

Golgi Apparatus

The **Golgi apparatus** looks like a stack of hollow curled pancakes surrounded by bubbles. Vesicles that depart from the endoplasmic reticulum fuse with the Golgi apparatus, delivering proteins and phospholipids to it. Once in the Golgi apparatus, the proteins and lipids are modified by the addition of sugars and phosphates (processes called **glycosylation** and **phosphorylation**, respectively). From here, the Golgi apparatus packages these molecules into new vesicles and sends them off to the cell membrane to be secreted or sends them into the cytoplasm to become part of the cell's population of vesicles, such as **lysosomes** (see below).

The Golgi apparatus with accompanying vesicles

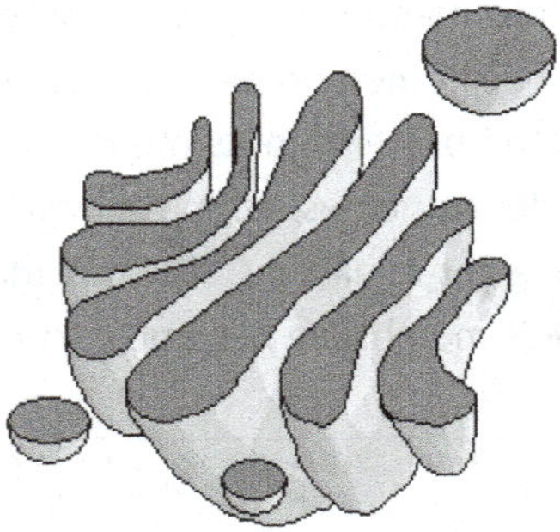

In the following image, we see the route that a molecule such as a protein would take to be either secreted from the cell or packaged into a vesicle like a lysosome. The route that a protein takes depends on the primary structure of the protein in question. Proteins destined for secretion typically have characteristic amino acid sequences that cause the Golgi to package it into secretion vesicles. Likewise, lysosome-bound proteins have characteristic amino acid sequences that the Golgi recognizes and uses to direct into newly forming lysosomes. Vesicles, such as lysosomes, depart the Golgi apparatus just like vesicles leave the endoplasmic reticulum: they bubble-up and detach.

A protein's path through the Golgi apparatus

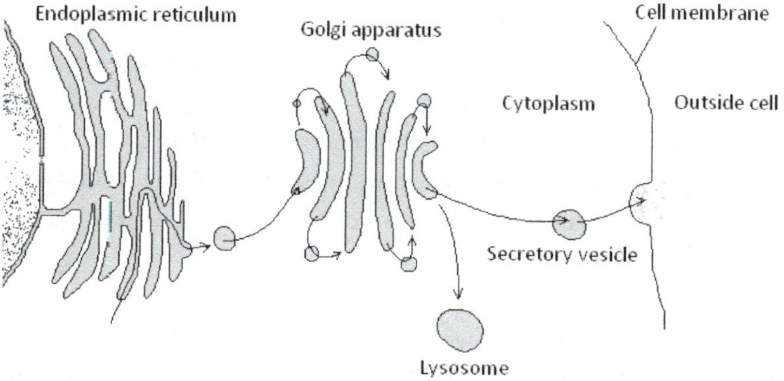

Lysosomes

Like the many vesicles you would see in a cell, **lysosomes** look like little bubbles. They are made by the Golgi Apparatus, and their job is to be the waste control specialists. Lysosomes contain enzymes that break down other proteins and chemicals, old cellular debris, or bacteria and viruses that have been phagocytized. The lumen of a lysosome is very acidic (i.e. low pH/high H^+ concentration). This acidic environment helps the enzymes inside the lysosome do their job to break down materials.

Whenever the lysosome is going to break something down, it fuses its membrane with the membrane of the target, usually another vesicle or worn-out mitochondrion or chloroplast. Once the membranes fuse, the contents of the lysosome mix with the contents of the target. The target is slowly digested. The basic chemicals that result from the breakdown of the target are usually transported out of the lysosome and into the cytoplasm for the cell to reuse, or are ejected from the cell when the lysosome fuses with the cell's membrane. The lysosome's enzymes are neutralized before this happens, however.

Another important use for lysosomes is in a process called **apoptosis**, or cell suicide. Basically, the lysosome is directed at the cell itself, digesting its critical structures and causing the cell to die. This may seem sad and tragic, but the health of the overall body depends on some cells removing themselves from the population. If the number of cells in the body cannot be controlled, cancer can result.

Lysosomes are generally only present in animals. Plants and fungi use their vacuoles to take care of waste products. Bacteria do not have lysosomes.

Peroxisome

Peroxisomes are vesicles that are found in nearly all eukaryotes. They play an important role in the metabolism of some long fatty acids, less common amino acids and a few carbohydrates. The telltale feature of peroxisomes is that they use peroxide during the course of their activity.

Endosome

One final vesicle we'll cover in the endomembrane system is the **endosome**. An endosome, whose name means roughly "in-body," is a little membrane bubble that results when something is engulfed during endocytosis. If you'll recall, endocytosis is the process of the membrane folding inward to make a bubble that is full of whatever was outside the cell. That bubble is called the endosome. Endosome are typically found near the surface of cells that undergo endocytosis or phagocytosis. They contain extracellular debris, and they are eventually joined with lysosomes, whose enzymes and acid digest the material inside the endosome.

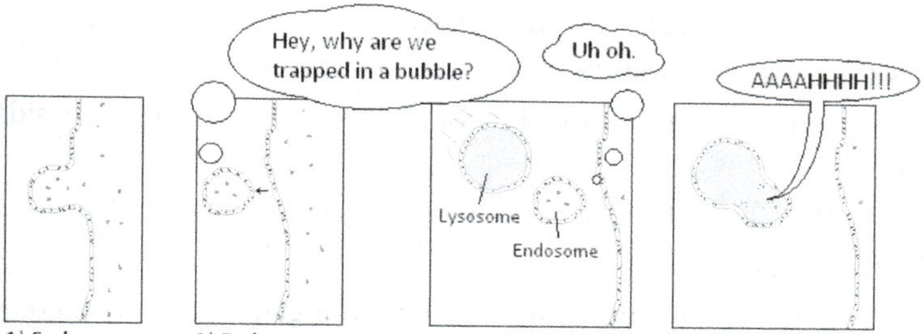

1) Endosome forming 2) Endosome moves inward 3) Lysosome swoops in 4) Lysosome merges with endosome. Particles inside endosome break down

The Symbiotic Organelles

The word **symbiotic** means that two organisms live in harmony, helping each other to thrive. There are two organelles within the cell that are believed to be symbiotic to eukaryotic cells. These are the mitochondria (singular mitochondrion) and the chloroplast. The idea is that long ago these organelles were bacteria-like organisms that originally preyed upon eukaryotic cells, perhaps invading eukaryotic cells from time to time to take advantage of the resources inside. Over great lengths of time (millions and millions of years), this relationship gradually shifted so that both the cell and the invading organism benefited from the interaction. The mitochondrion and chloroplast are both involved in energy management, and so the eukaryotes that harbored them gained the benefit of more efficient energy handling. By offering their services of energy management, the ancient mitochondria and chloroplasts no longer had to live the rough and rowdy life of a cellular pirate.

The reason this theory was generated is that both the mitochondrion and the chloroplast have their own DNA. In order for living organisms to survive and reproduce on their own, they must have DNA, suggesting that once upon a time mitochondria and chloroplasts lived on their own. If so, their DNA is mostly out-of-practice now, however, as it would have been millions and millions of years since they lived on their own, but the DNA they have still serves as the blueprint for many (though not all) of the proteins found inside them.

That said, let's look at these two organelles more in depth.

Mitochondria

Mitochondria are membrane-bound organelles where a great deal of energy is harvested from several specific basic organic chemicals. They are, in a sense, power plants for the cell. Within their membrane is a second membrane, like a balloon within a balloon, only this inside membrane is ruffled, like a rug that is scrunched up.

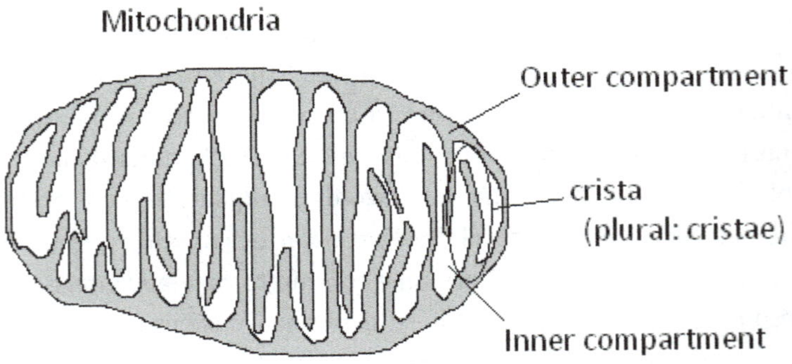

Within this inner ruffled membrane are the important proteins that function in the reactions that generate the energy for the cell. These particular reactions comprise the **Krebs Cycle** (also called the **citric acid cycle**, or the **tricarboxylic acid cycle**), which we will discuss in detail later in the book. For the moment, a brief description of that cycle is as follows: glucose (a sugar) is broken down into smaller pieces, and these pieces are eventually transported into the mitochondria where most of the energy contained in the molecules is extracted and pumped back into the cell. In addition to this, mitochondria also play a role in the metabolism of fats.

New mitochondria are made when other mitochondria divide, much like a bacteria does.

Chloroplasts

Chloroplasts are membrane-bound organelles found in plant cells and look like bags that contains connected stacks of tightly-packed hollow pancakes (called **granums**). Chloroplasts can be thought of as a plant's solar panels. They are responsible for capturing energy from the sun in a process called **photosynthesis** (which we will discuss in an upcoming unit). Chloroplasts are full of green pigments. They are the reason leaves are green. The details of their anatomy are shown here:

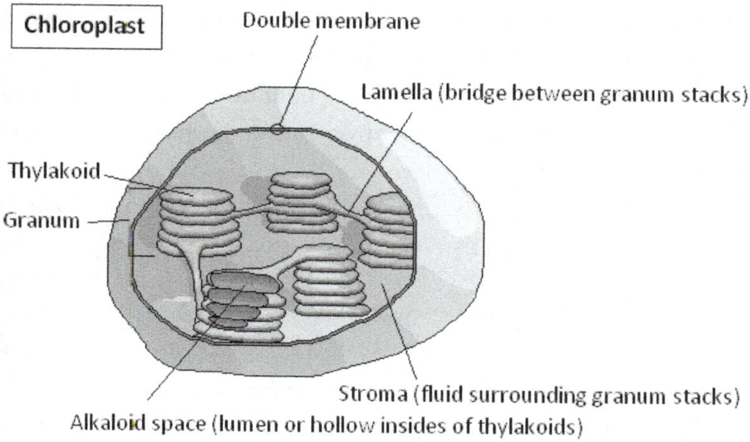

Like mitochondria, chloroplasts have their own DNA, but many of the genes responsible for the chloroplast's upkeep are now in the cell's nucleus. Chloroplasts divide like bacteria do as well.

The Cytoskeleton

The **cytoskeleton** is a system of long fibrous proteins that crisscross the cell. Their general role is to provide structure and support for the cell and its many other organelles, although some fibers serve to assist in cellular movement and material transportation as well. There are three broad categories of protein fibers in the cytoskeleton: the microfilaments, the intermediate filaments, and the microtubules. Refer to the large image near the end of this section for a visual description. We'll look at each individually.

Microfilaments

The **microfilaments** are the smallest of the three categories of protein fibers in the cytoskeleton. The best example of a microfilament is actin, which we encountered before in our study of belt desmosomes. These tiny filaments are made of monomers of actin (called G-actin) that come together to make a long polyprotein (meaning "many protein") polymer. Curiously enough, the actin monomers tend to add to the fiber from one end, while they leave the fiber from the other end.

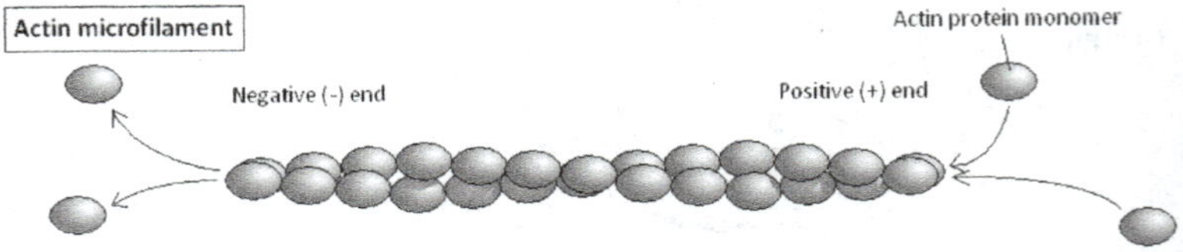

The peculiar nature of how the monomers combine to make actin gives these fibers a polarity of sorts, (although not in the chemical sense that we talked about earlier in the first topic text). This polarity allows other proteins that interact with the actin to "know" direction. It's like finding a road in the wilderness that has arrows painted on it, directing you to the nearest sign of civilization. . . which coincidently would be the farthest spot from my apartment.

Similarly, various transportation proteins (usually from a family of proteins known as **myosins**) find the actin fiber and orientate toward or away from the positive end of the actin. These proteins, in a sense, walk along the actin in one way or the other, carrying materials across the cell to wherever they need to go. One type of protein might attach and always travel towards the positive end, while another type of protein might attach and always travel towards the negative end.

Actin is also very important in muscles. Basically, myosin proteins are lined up with actin filaments, and when the signal is given all the myosin proteins pull on the actin filaments to cause the cell to shorten. It's like a team of people playing tug of war with a wall – they just pull themselves toward the wall. Here is a very simplified version of what happens:[16]

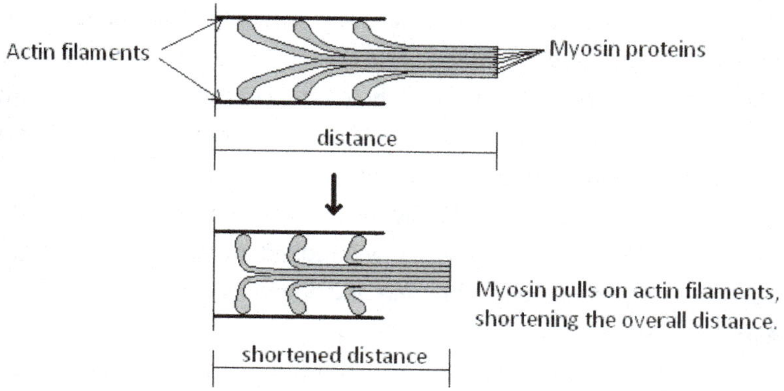

Depending on the type of cell, actin is found in highest density around the inside of the cell's membrane. There it sometimes

[16] For those interested, the image shown is only half of a structure called a sarcomere, which is covered more in depth during the second semester of biology or in courses on anatomy and physiology. I'll be talking more about it in future topic texts.

helps to move the cell membrane (by a mechanism similar in principle as muscle) to assist in endocytosis (i.e. engulfing material).

Intermediate Filaments

Intermediate filaments are larger in size than microfilaments and have a more complex structure. The basic unit of intermediate filaments contains four short protein threads that come together to make a pair of dimers (dimer = two monomers). Since there is a *pair* of dimers, we call the basic unit a tetramer (tetramer = four monomers). These tetramers are lined up back-to-back-to-back in very long threads (See next image).

The primary purpose of intermediate filaments is structure and support. Their design tends to make them rather elastic in nature, which affords them the ability to survive deformations. There are lots of different kinds of intermediate filaments, and they are found in a variety of places. A common example is a family of filaments called **lamins**. These form a network just on the inside of the cell's inner nuclear membrane. This network helps to give the nucleus its round shape. Without this scaffolding, the nucleus would just be a flimsy squishy bag flapping around in the cytoplasm.

Another common family of intermediate filaments is the keratins. We saw these when we looked at spot desmosomes. Keratin is very common in hair, nails and animal horns.

Microtubules

Microtubules are the largest of the protein fibers in the cytoskeleton. Their basic unit is a pair of protein blobs called tubulin. One is alpha tubulin and the other is beta tubulin. This dimer of protein blobs lines up back-to-back-to-back with other alpha-beta tubulin pairs to make long threads (called **protofilaments**, see next image). Thirteen of these protofilaments line up next to each other to make a tube. Like actin, microtubules also have directionality to the way they fit together. The negative end is usually anchored somewhere near the center of the cell, while the positive end continues to grow outward by the addition of more alpha-beta tubulin dimers. Also like actin, this directionality gives microtubules an important role in transportation of cell material. For example, when a protein called **dynein** binds to the microtubule, it will begin to "walk" toward the negative end, and in doing so it carries materials with it toward the cell's center. Another protein, called **kinesin**, will bind to microtubules and walk toward the positive end, carrying cell material *away* from the cell's center.

Microtubules play important roles in cell division, as we will see in future topic texts.

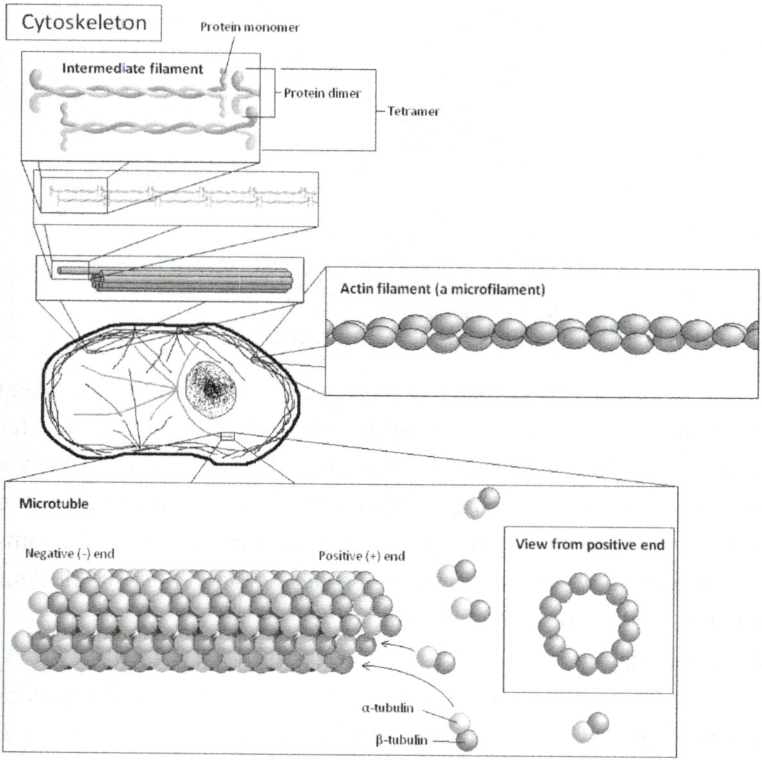

Centrioles

The **centriole** is a barrel-shaped structure that is very large compared to microfilaments, intermediate filaments and microtubules. Centrioles are usually found in pairs positioned at roughly 90 degrees from each other, an arrangement called a **centrosome**. Centrioles are made of lots of microtubules. The general design is a set of nine bundles of microtubule **triplets** arranged in a star-like fashion.

Two Centrioles Arranged into a Centrosome

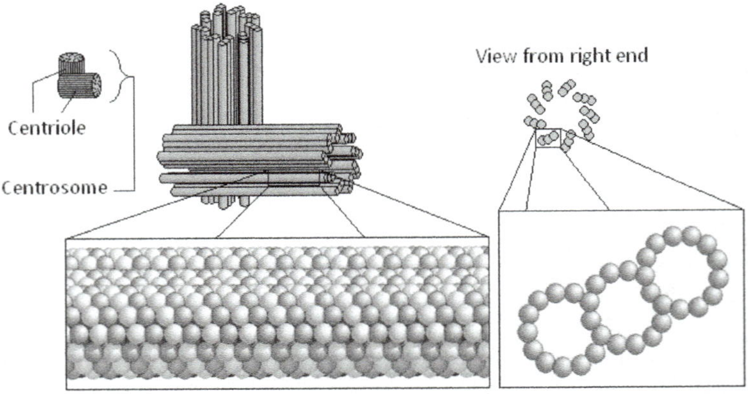

Theories concerning the purpose of centrioles have changed over time, but they are presently believed to be important for determining the position of the nucleus. Also, they seem to have a clear role in the positioning and functioning of cilia and flagella (see below). Although centrioles are not always drawn in simple diagrams of the cell, when they are, they are usually depicted near the nucleus when the cell is preparing to divide.

Cilia and Flagella

Cilia are small protein-containing hairs that cover the surface of some kinds of eukaryotic cells. These hairs wave back and forth to cause one of two different effects. 1) If the cell is stuck in place, the cilia force water to flow in one direction, which is important in many organ systems such as the lungs and female mammalian reproductive tract. 2) If the cell is mobile, the cilia help to move the cell around, like oars on a ship.

Cilia have a centriole-like microtubule system called an **axoneme** in the center of their column. The axoneme is shaped like a rim of microtubule-fused pairs (**doublets**) around the outside, with two independent microtubules on the inside (see top-right part of the following image). Between each of the doublets are dynein arms. If you recall from the discussion about microtubules above, dynein tends to grab hold of microtubules and walk in one direction. The dynein in cilia sprout from one doublet to grab the next. As they do this, the dynein try to "walk" up the nearby microtubule, causing the entire structure to bend. All of the dynein arms in the cilia are designed to work in an orchestrated fashion with each other so cilia bend this way and that, waving back and forth.

Since axonemes have nine fused microtubule pairs around the rim and two independent microtubules on the inside, the structure is typically dubbed the **9 + 2** structure. At the very base of the cilia's axoneme, just inside the cell's membrane, is a **basal body**, which is a structure that is nearly identical to (and derived from) the centriole.

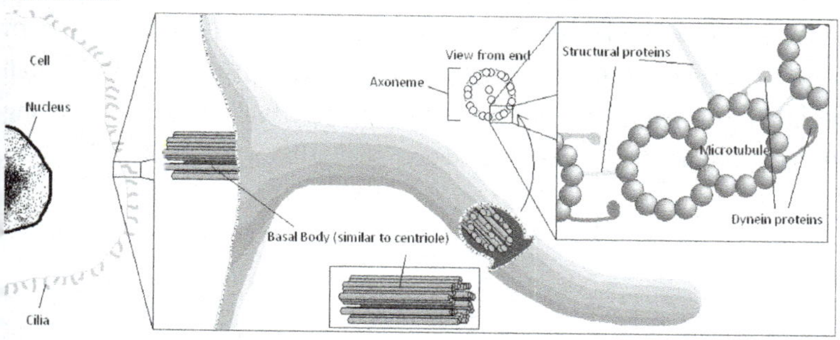

Flagella are longer than cilia and are used primarily for locomotion. The flagella found in eukaryotes, such as in sperm, have a structure that is essentially identical to cilia, only longer. The flagella in prokaryotes, however, are structurally complex and very different. We will not go in depth into that topic as it is outside the scope of this book, suffice it to say that prokaryotic flagella work by spinning around their axis (like a propeller pushing a boat through water), while eukaryotic flagella wave back and forth (like an oar or a fish's tail).

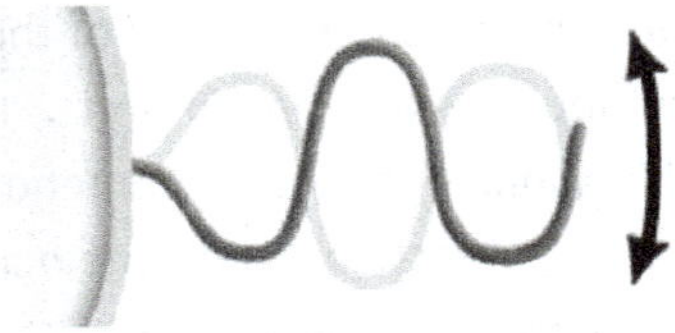

Eukrayotic flagella motion:
wavelike, back-and-forth

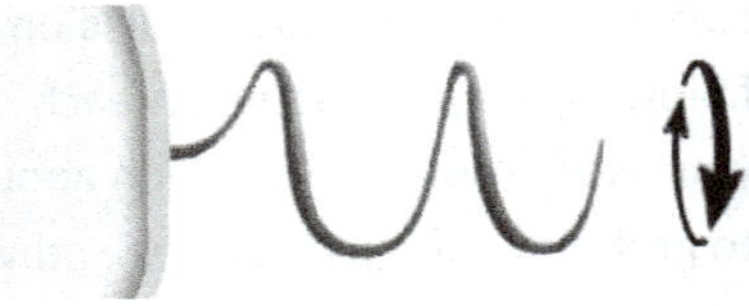

Prokaryotic flagella motion:
rotating

One final note should be made about the cilia and flagella. In previous sections, we talked about the glycocalyx that many cells have. The glycocalyx, as you'll recall, is made from small carbohydrate molecules attached to proteins and lipids in the membrane. This creates a very tiny fuzzy covering attached to the membrane. Do not confuse this with the hair-like cilia or flagella. Cilia and flagella are MUCH larger than the glycocalyx. You would have a hard time seeing anything like the glycocalyx with any equipment short of an electron microscope, whereas structures like cilia and flagella can be seen with light microscopes (and the appropriate dye). Cilia are still small enough, however, that you would probably have a hard time seeing them in the light microscope as well.

Summary

In this section, we talked about the anatomy of the cell.

- We divide cells into two broad categories: eukaryotes and prokaryotes. Eukaryotic cells have nuclei, while prokaryotic cells do not.
- Cells are made of organelles. These include: the nucleus, nucleolus, chromosomes, ribosomes, endoplasmic reticulum (rough and smooth), Golgi apparatus, lysosomes, peroxisomes, mitochondria, chloroplasts, microfilaments, intermediate filaments, microtubules, centrioles, cilia, and flagella.
- The nucleus is the headquarters of the cell. It houses the DNA.
- A chromosome is a complete polymer of DNA coiled up around protein complexes called nucleosomes. These are, in turn, tightly wound together in a very compact manner. Chromosomes are most compactly wound during the process of cell division.
- The nucleolus is the part of the nucleus that contains unwound DNA, proteins and RNA which are involved in the initial stages of the protein production. This is where the "blueprints" for proteins are made.
- The nucleus is surrounded by a double membrane. This double membrane is continuous with the endoplasmic reticulum. The nuclear membrane has numerous openings that are guarded by nuclear pore protein complexes. These complexes facilitate and regulate the traffic between the cytoplasm and nucleus.
- Ribosomes are large protein/RNA complexes. The RNA in ribosomes is called ribosomal RNA (or rRNA). Ribosomes

have a large and small subunit that fit together like two gloves.

- Ribosomes make protein by receiving blueprints in the form of RNA (called messenger RNA, or mRNA). This RNA threads through a channel in the ribosome. The new protein emerges from a small tunnel in the backside of the ribosome.
- The **endoplasmic reticulum (ER)** is a series of flattened tubes directly attached to the nuclear membrane and is part of the endomembrane system of the cell. There are two kinds of endoplasmic reticulum: the smooth endoplasmic reticulum and rough endoplasmic reticulum. The rough endoplasmic reticulum has ribosomes attached to it. These particular ribosomes thread the proteins they are making through a small channel in the endoplasmic reticulum, so that the proteins ultimately end up in the space inside the endoplasmic reticulum. This space is called the lumen, a word that means roughly "hollow space within." Alternately, ribosomes may embed new proteins into the membrane of the ER.
- One of the most important roles of the ER is to manufacture triglycerides. These triglycerides, as well as any proteins embedded in the membrane, get packaged into small vesicles that depart from the ER toward the Golgi apparatus.
- The Golgi apparatus looks like a stack of hollow pancakes with bubbles around it. Vesicles merge with the Golgi from the ER. Once in the Golgi, the transported material gets chemically modified. When complete, the Golgi releases a vesicle (with chemicals inside) that goes to either the cell membrane or toward the system of lysosomes in the cytoplasm. The Golgi also makes lysosomes in this fashion.
- Vesicles that merge with the cell membrane release their contents to the outside of the cell.

- Lysosomes are vesicles that contain acid and enzymes that breakdown cellular debris and engulfed microorganisms.
- Peroxisomes are vesicles that metabolize long chain fatty acids, uncommonly found amino acids and some sugars. They use peroxide to metabolize many of these chemicals.
- Mitochondria harvest energy from various basic biological molecules. Mitochondria are believed to have once been free-living organisms. They have their own DNA.
- Chloroplasts are found in plants. They contain chlorophyll (which is what makes plants green). The chlorophyll captures light and helps the chloroplast turn the light into chemical energy. This chemical energy is used by the rest of the plant.
- The cytoskeleton is a series of structures made from microfilaments, intermediate filaments, and microtubules. Centrioles, cilia, and flagella are included in this system.
- Microfilaments are the smallest elements of the cytoskeleton. Actin is the most commonly cited example. Actin is made of globular actin protein that connects together to form a double spiraling filament. Actin is used for structure, cellular transportation and cellular movement. Actin is found in muscle, where it takes an active role in muscle contraction.
- Intermediate filaments are intermediate in size to microfilaments and microtubules. They are made from threadlike proteins that pair up. These pairs combine with other pairs to make tetramers. The tetramers line up to form very long threads. Intermediate filaments tend to be elastic in nature. They are commonly found in hair, nails and horns.
- Microtubules are the largest of the cytoskeletal filaments. They are made of pairs of alpha and beta tubulin that line up into protofilaments. Thirteen protofilaments arrange in a circular tube fashion. Microtubules serve structural roles as well as roles in cell material transportation and movement.

- Centrioles are large barrel-shaped structures (usually found as pairs of barrels positioned at roughly 90 degrees from each other). Centrioles are believed to help determine the position of the cell nucleus and are important in the functioning of cilia and flagella.
- Cilia and Flagella are hair-like protein structures found in the cell's membrane. Cilia are smaller. They have two roles: 1) they make water flow over stationary cells, and 2) they move non-stationary cells around. Flagella are much larger and are used to move the cell around. Eukaryotic flagella have the same structure as cilia but are longer. Prokaryotic flagella have a different (more complex) structure than eukaryotic flagella.

Unit 7

Energy and Photosynthesis

In this section we cover:

1) The nature of energy
2) The sun, light, and electromagnetic radiation
3) Photosynthesis
4) The Calvin Cycle

In the previous topic text we learned about how particles come together to make atoms, how atoms come together to make molecules, and how molecules come together in the reactions of life to make larger molecules such as DNA, polysaccharides and fatty acids. Then in this book we studied cell structure and function. In this section, we begin our discussion of energy and how it affects life.

Energy

The precise definition of energy requires a great deal of confusing philosophy, but the definition has to do with the ability of one system to do "work" on another. For most novice students, this jargon brings up images of time cards and annoying bosses ordering people around, so let's instead try a makeshift definition to get our feet wet. By approximation, *energy is the ability of one thing to change another thing*[17]. If one object has no energy, it will be unable to alter another object. Similarly, if the first object has a lot of energy, it has the potential to cause great change to another object. The Universe is a complex thing, however, and the ways in which one object can change another depends on the circumstances and the

---

[17] For example, one energetic object might change another object's speed, temperature, structure, appearance, etc.

types of energy present. Scientists recognize many forms of energy, such as thermal energy (heat), light, sound, chemical, nuclear, etc. Let's look at a few examples to get some practice with the concept.

Consider heat energy. If one object has more heat energy than another object, it can change the second object by heating it up, *changing* its temperature. The change isn't necessarily permanent. If you remove the hot object, the second object will cool down again as the heat dissipates into the air. The change isn't always obvious either. If you put a battery (chemical energy) into an electronic device, the battery will begin to cause microscopic changes by forcing electrons to move around inside the device.[18] Naturally, you would not notice this form of change unless the device had some kind of output or observable behavior, such as making sound, light, heat, etc. For example, the sound energy made by the device briefly *changes* the positions of little hairs in your ears. This starts a chain reaction that alerts your brain to the sound. As another example, the light energy from the electronic device briefly *changes* tiny molecules in your eyes. Once again, this starts a chain reaction that ultimately alerts your brain of the light energy. To summarize, the energy from the battery gets transferred to the electronic device, which then causes changes inside your body.

Battery-powered devices are not the only things changing your body, however, and you can't always tell when it happens. In fact, energy that brings change that is hard to detect can be very dangerous. For example, in the first topic text we talked briefly about what radioactivity is. Radioactivity results from nuclear energy. Nuclear energy causes changes in the nuclei of atoms, which eject radioactivity into the environment. This radioactivity can cause tiny changes inside your body, such as in your DNA. If the change occurs at the wrong place, it can lead to cancer. Nuclear energy isn't the only

---

[18] The movement of electrons is basically what electricity is.

form of energy changing your body though. As we saw with the electronic device, other forms of energy are constantly changing your insides . . . heat, vibrations, sound, electricity, chemical. . . They are attacking you all the time. Sometimes it's subtle. At other times it's obvious and even violent. With all the changes happening in your body, you are faced with the constant threat of illness and death. The only solution, however, is to fight fire with fire, as they say. We must use energy to reverse the changes. We require energy to thrive. If we do not have it, we cannot correct the changes taking place within us or around us. Our insides and outsides will gradually become less suitable for life, and then we'll die.

Though we are surrounded by energy, we can't use just any ol' type. Our bodies require specific forms, primarily that of food and drink. We must eat to both minimize harmful change and encourage healthy change. It is a fact of life. It's like the old saying: there are only four constants in life: taxes, death, our need to eat and loud neighbors. Energy powers it all.

Energy and Sun

Let's take a step back to get a grasp on where all this starts. The energy that biological systems use begins with the sun. The sun is a big ball of nuclear energy. That energy is constantly rearranging the sun's atomic nuclei, and this emits massive amounts of radioactivity. Some of that radioactivity is called **electromagnetic radiation**, or EM radiation, which is just a fancy name for a whole range of particles of light (**photons**) that have different energy values. The details are vast and complex, but basically EM radiation travels as waves and comes in a very large spectrum, from the extremely energetic **gamma rays** to the relatively weak **radio waves**. Between the two, within an intermediate range of energy, is *visible* light. Consider the analogy of rain. Some drops are large, while others are small. EM radiation is similar. Some photon particles have incredibly high amounts of energy, while others have less. All of these particles of light are racing past us all the time, but more so during the day. Although it is difficult to show in grayscale, the picture below should give you an idea of the spectrum of electromagnetic radiation.

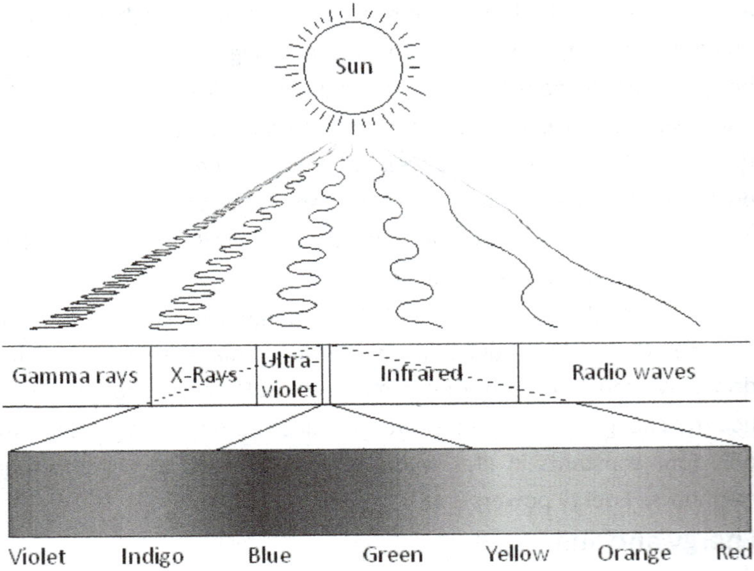

Collectively, EM radiation represents a *lot* of energy hitting the Earth at any given instant. A long time ago, living systems found a way to use that energy to power the changes they needed to survive: they developed the ability to absorb the energy that was in light. Up until then, living systems had to scavenge around for whatever energy they could find in their inorganic environment. With the advent of harvesting energy from the sun, things began to change radically. These light-absorbing organisms began storing the energy they caught from the sun's rays as highly energetic molecules. The scheme worked very well. The harvest was so bountiful, in fact, that energetic molecules began leaking from these organisms into the environment (and still do). One of those highly energetic molecules is oxygen, O_2.

Not all organisms had developed the ability to absorb light, however, but the increasing presence of oxygen gave these latter individuals a new source of energy as well. Life all across the planet took a huge leap forward as a clear distinction emerged between two lifestyles that are still in existence today. The first lifestyle is that of the **autotrophs**, which are the organisms that get their energy directly from a primary source (i.e., sun or inorganic energy source). Modern day examples include plants and algae. The second lifestyle is that of the **heterotrophs**. They get their energy by eating autotrophs and/or their byproducts. In effect, the autotrophs pull inorganic energy into the reactions of life, and the heterotrophs freeload off of them. Modern examples of the freeloading heterotrophs include herbivores, carnivores and politicians.

The important point to understand is that nearly all the energy that fuels life starts with the sun, gets captured by autotrophs and then gets used by heterotrophs. Even if one heterotroph makes a living by eating only other heterotrophs, (such as a carnivore), the energy that it gets still originated from the sun. No community of organisms can survive without autotrophs being part of the picture somewhere along the line.

The absorption of light energy and its storage within biological molecules is called **photosynthesis**. When autotrophs do photosynthesis, they take low-energy molecules (CO_2 and water)[19] and

---

[19] CO_2, H_2O, and N_2 are actually very common low-energy molecules found on planets having atmospheres. The reason is that carbon, oxygen, hydrogen, and nitrogen are common atoms, and when they form these molecules, they are very stable (e.g. happy in that form). The Earth is unique in that it has a LOT of O_2 in the atmosphere because of photosynthesis. O_2 is very energetic (e.g. unstable or unhappy). If plants died off, our atmosphere would quickly return to something similar to the other planets. In that regard, photosynthesis may be seen as one of the most important transforming agents of our planet. Therein lay one of the reasons for the great concern over deforestation.

use energy to combine them in a series of reactions to make sugar. The overall reaction goes as follows:

$$6CO_2 + 12H_2O + \text{light} \text{ -----> } 6O_2 + C_6H_{12}O_6 + 6H_2O$$

In other words, six carbon dioxides, twelve waters and the energy from light are mixed together in a tiny biological pot to make one sugar (specifically **glucose**, $C_6H_{12}O_6$). In addition to the glucose, six oxygen molecules and six of the original twelve water molecules pop out. This is how plants make oxygen to fill up the environment.

So why would a plant do this? What is the advantage of having a sugar sitting around? Well, the sun isn't there 24 hours a day. Cloudy days happen too. Sugar is highly energetic and represents the plant's attempt to save for a raining day. At night and during other low-light conditions, the plant cannot rely on photosynthesis to drive its reactions, so it uses the sugars it made earlier to continue living. All the plant needs to do is recombine the sugar with oxygen and water, run the reaction in reverse, and *TADA* the energy pops back out, only this time the energy is *not* in the form of light; it's chemical energy. This energy powers the plant. We use sugar in the same way. We breathe in oxygen and drink water, and combine these things with high energy molecules like sugar that we've eaten to acquire the energy we use to drive the reactions we need to keep us alive.

At this point, I should point out a very important fact. Plants do not release oxygen for our benefit. They don't do it because they want to be nice. They release oxygen because they can't hold it all in after it is produced. Plants use oxygen just like we do. It is useful to them, but they create so much of the stuff, that there isn't a need to be stingy. Not only that, but excess oxygen can interfere with the synthesis of glucose. Too much oxygen can be a bad thing. So, no, plants aren't being friendly to us.

Photosynthesis: The Light Reactions

Photosynthesis starts with the "light reactions," so named because it requires light to drive it. Light enters the plant's many chloroplasts (remember that from the last unit?) and hits molecules called **chlorophyll**. There are many kinds of chlorophyll actually, but the general form is shown in the picture below. The structure looks like a little satellite dish, with a single magnesium ion stuck in the middle, locked there, with dreams and ambitions, plans for greatness, all ruined by the wickedness of the chlorophyll. Chlorophyll looks green, so that means it must like green light, right? Sadly, no. Chlorophyll absorbs visible light mostly in the range of violets, blues, and reds. The result is that the green light bounces off unabsorbed. This is the reflected light that we end up seeing, making leaves look green.

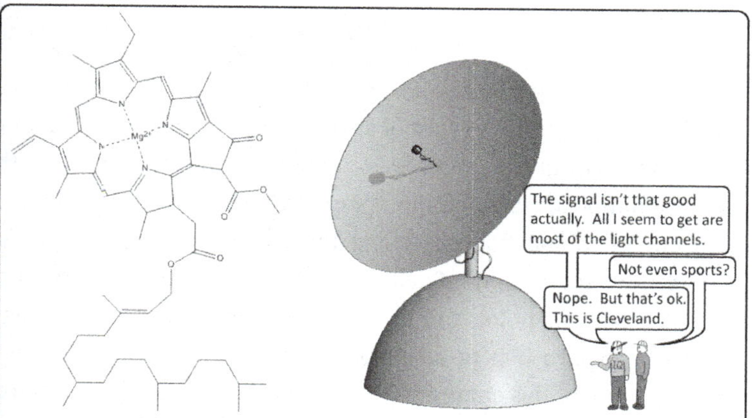

Chlorophyll molecules look like satellite dishes.

As it turns out, chlorophyll isn't the only molecule working in the craft of absorbing light. Chlorophylls have other molecules called **carotenoids** that help in the process. Together with these accessory pigments, the chlorophyll and carotenoids make up what's known as a **light harvesting complex**, (**LHC**). A bunch of these magnesium-mistreating baddies are bunched together to make a **photosynthetic unit**. Within each unit, the numerous light harvesting complexes gather up energy from the light and transfer it to a centralized area of the photosynthetic unit for the next step in the process. This centralized area is known as the **reaction center**.

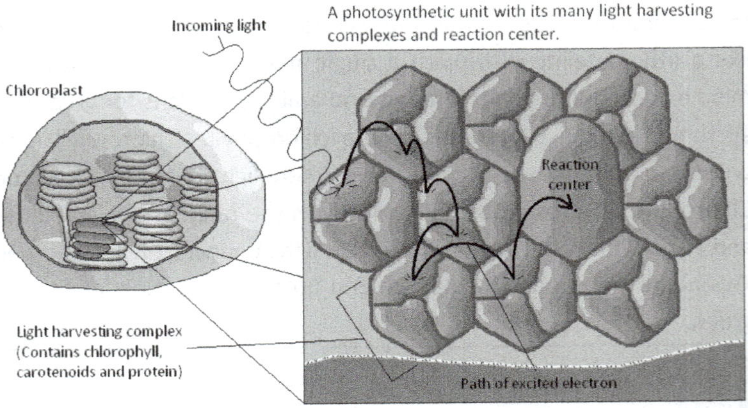

This last image also gives a simplified take on how light is absorbed. In essence, the light of specific energy comes in and bumps an electron from its molecular orbital shell from within the magnesium-holding part of the chlorophyll. This electron becomes quite energetic and bounces along the pigments like a hot potato being passed around, until finally the electron arrives at the reaction center. From this point, the electron uses the reaction center as a launching point to hop to a number of other molecules, each jump releasing some of this energy in a controlled manner until the electron finally arrives onto a molecule called **$NADP^+$**. This entire event happens twice so that two electrons collect on the $NADP^+$. Electrons have negative charges, so the two electrons transform $NADP^+$ into $NADP^-$. This minus charge attracts H^+, which combines to make it **NADPH**. NADPH then floats away. It has more energy than just regular $NADP^+$, so it takes this energy away to other parts of the cell. The following image illustrates the concept. Here, the photosynthetic unit is designated "P700," a name that refers to the wavelength of light that excites it. The names of a few other proteins are included as well.

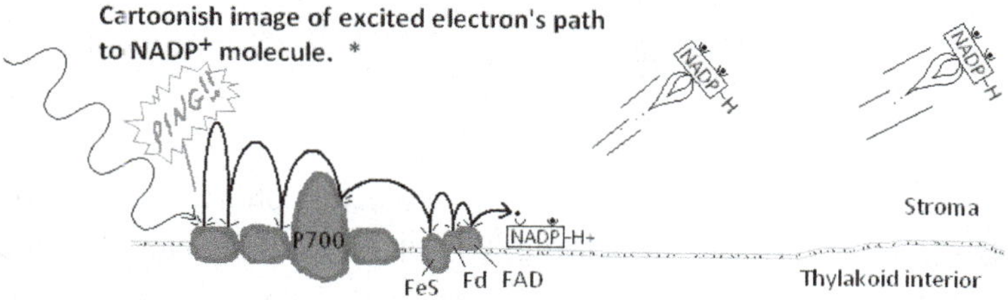

* In reality, the electron does not actually bounce along the membrane structures, but rather travels *through* the structures. Depicting the electron as a bouncing ball makes the process easier to remember though.

The story of the chlorophylls isn't quiet over just yet, however. For each electron that gets excited by light (bouncing around and off the photosynthetic unit and onto NADP$^+$), there is a hole left behind where the electron was.

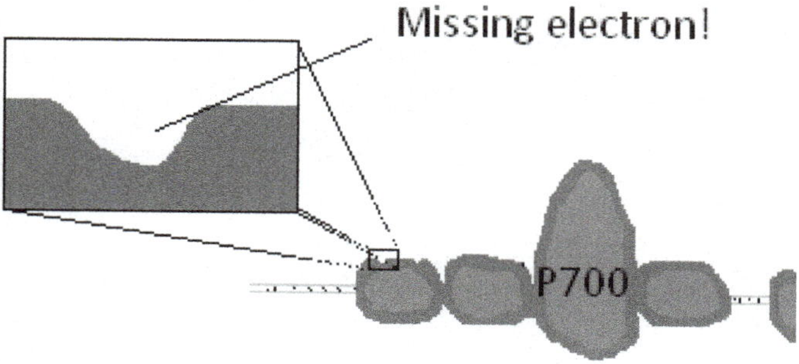

These holes need to be filled, or else the process will eventually come to a halt! To help fill these holes, modern plants use excited electrons from a *second* photosynthetic unit (called P680) to fill the first. This second photosynthetic unit is slightly different than the first one, because it requires slightly more energetic light to excite electrons. Otherwise, it works pretty much the same way, with electrons moving from site to site as in the following picture:

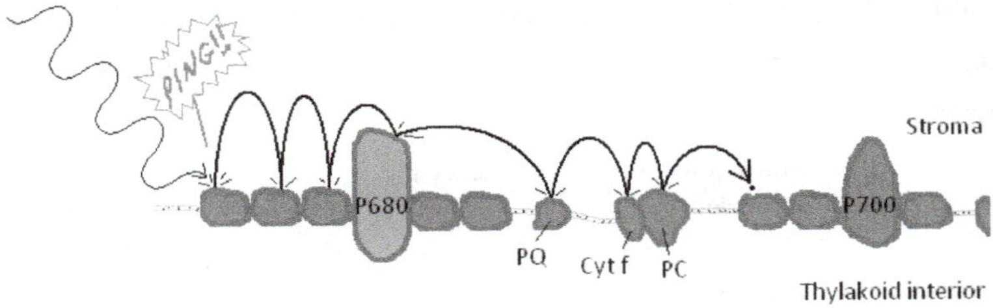

This, however, leaves the second photosynthetic unit deprived of electrons. To solve the problem, the P680 complex steals electrons from water, one for each time light bounces its own electrons down the line toward $NADP^+$. Ultimately, this causes water to split up into one oxygen atom and two protons, as shown to the here:

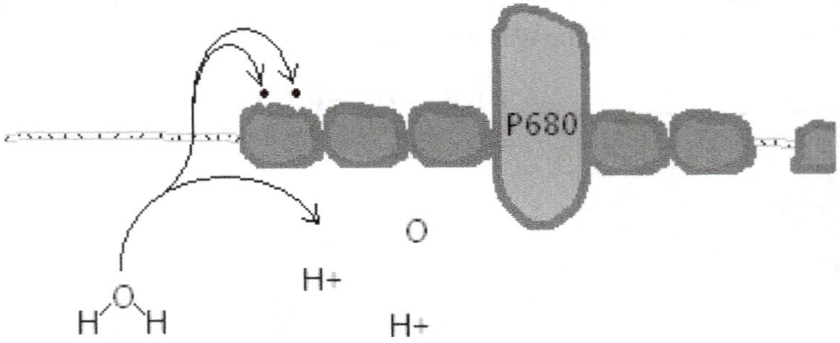

The lone oxygen that is created in the process brings us back to some of the basic chemistry we talked about in the first topic text. Lone oxygen atoms are not very happy because they do not have a full valence shell. Luckily, the photosynthesis process happens at an incredibly fast speed so that the lone oxygen does not have to wait long before another oxygen atom is created. The two oxygen atoms pair up to make molecular oxygen,[20] O_2. Plants are not airtight systems, however, so some of the O_2 that is created escapes into the air. This innocent little O_2 leak is the reason our atmosphere is so radically different than other planets.

The story of photosynthesis isn't quite over, though. After all, it wasn't necessarily the plants' goal to make O_2. Plants do photosynthesis to capture the sun's energy in chemical form, so we need to cover a few more details. Well, in addition to pushing electrons onto an $NADP^+$ molecule, (affectively charging the molecule with energy), the photosynthesis process uses the energy of the excited electron to pump protons across the membrane. Putting it all together, it would look like this:

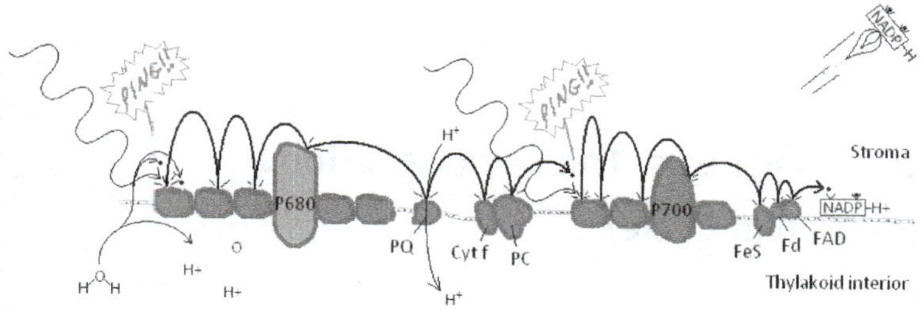

We are down to the homestretch. If you look at the last image, you'll notice that a bunch of H^+ ions are created/transported inside the thylakoid (the hollow pancakes inside the chloroplast) during the photosynthesis process. This leads to a very heavy overcrowding of H^+ in the thylakoid interior. If you recall from our discussion on concentration gradients, anytime you get an overcrowding of any chemical, that chemical will want to move in the direction that makes less crowding. That means the H^+ inside the

---

[20] As we'll see in the next unit, molecular oxygen, O_2, isn't very happy either, but the two atoms pair up to temporarily solve their problem of not having a full valence shell. This little marriage of convenience doesn't last very long.

thylakoid are unhappy and want out. This is where the plant uses a very creative trick: it doesn't let the H^+ out unless they pay a toll in the form of energy. That energy is used to convert ADP (a low-energy molecule) into ATP (a high-energy molecule) during a process called **phosphorylation**, which basically means "adding a phosphate." The highly energetic ATP floats away and helps power the rest of the plant.

Simplified chloroplast with crowded thylakoid interiors

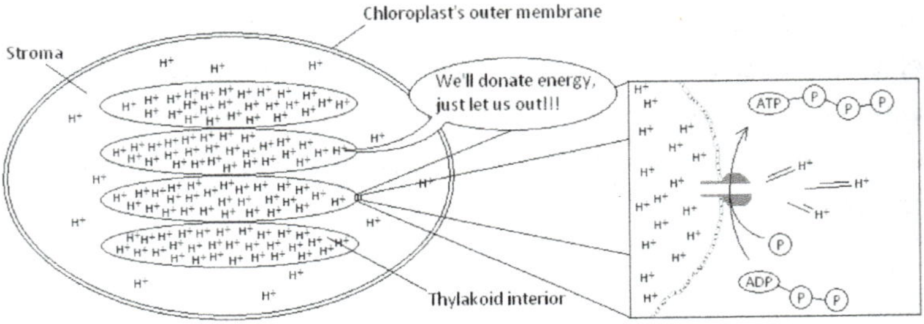

Photosynthesis: The Dark Reactions (The Calvin Cycle)

The next part of photosynthesis is referred to as the "Dark Reactions." This is not because they occur in the dark, but because they do not require light. They take place at any time. The general idea is that ATP and NADPH, (the two high-energy products from the previous set of reactions), are used to drive a cycle that puts together glucose. This is called the **Calvin Cycle**.

Step 1:

The first step in these reactions involves the combination of a CO_2 molecule with the five carbon sugar called **ribulose bisphosphate**, or **RuBi**. This makes a six carbon molecule that rapidly breaks into two **3-phosphoglycerate** molecules, shown next.

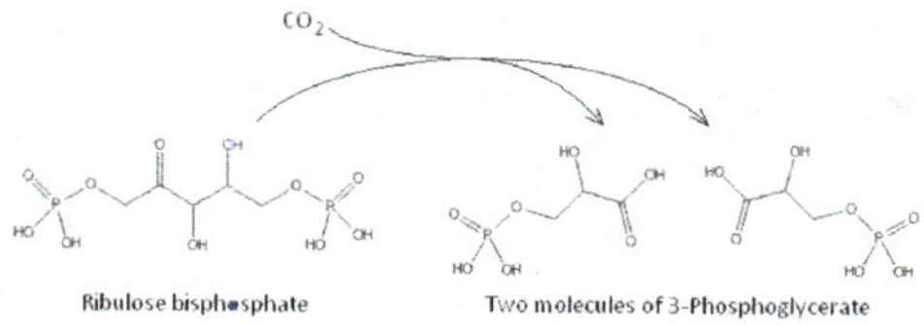

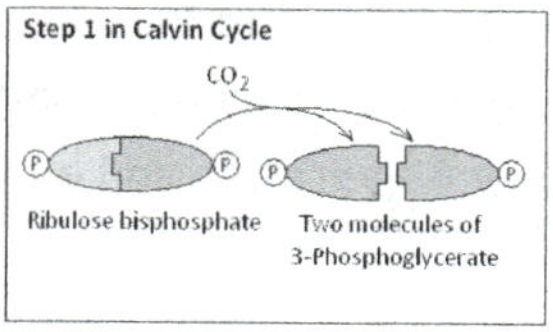

Step 2:

The next step in the cycle involves the phosphorylation of 3-phosphoglycerate by ATP, forming **1,3-bisphosphoglycerate** and ADP (reaction shown below). ATP is a very energetic molecule, so any time you see ATP being used (rather than produced) in a reaction, think "here is where energy is required to keep the process going." So, in this case, the principle tells us that converting 3-phosphoglycerate to 1,3-bisphosphoglycerate requires energy. This also means that 1,3-bisphosphoglycerate is more energetic than 3-phosphoglycerate. In the pictures that follow, I draw little "charged-battery" symbols on the molecules to give you an idea of how much energy they have relative to each other.

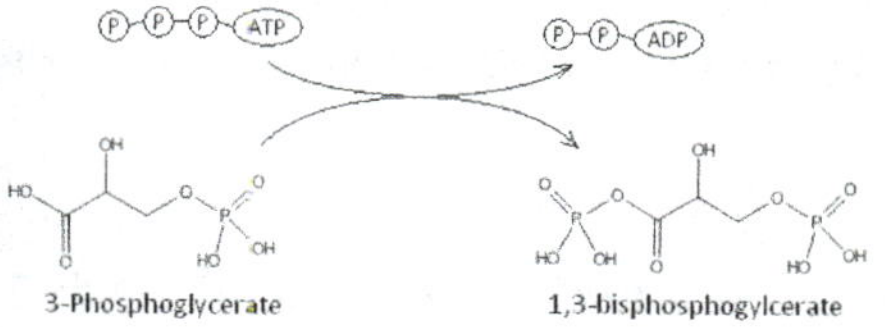

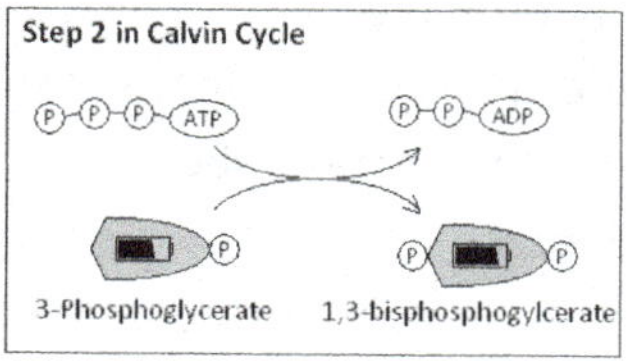

Step 3:

After the creation of the energetic 1,3-bisphosphoglycerate molecule, a molecule of NADPH swoops in to convert it into **glyceraldehyde 3-phosphate**. Like ATP, NADPH is a highly energetic molecule made during the light reactions of photosynthesis. Also like ATP, any reaction that involves NADPH is a reaction that requires energy input. Below is an image of this reaction. Notice from it that NADPH has a hydrogen when it begins the reaction, but none as it floats away when the reaction is done. The "missing" hydrogen atom was attached to the glyceraldehyde 3-phosphate molecule. That leads us to another general pattern in these kinds of reactions: NADPH gives a hydrogen atom and energy to the molecules it reacts with.

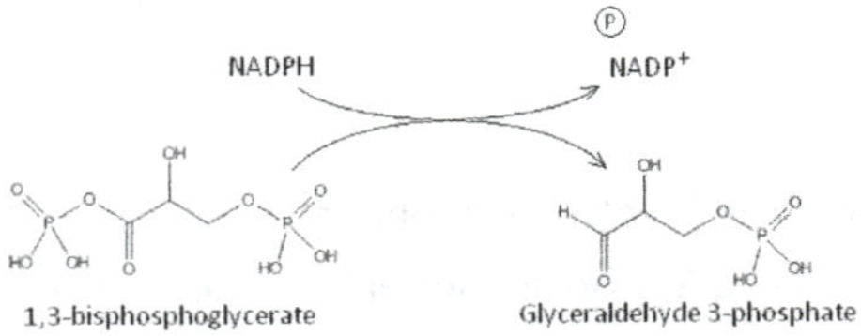

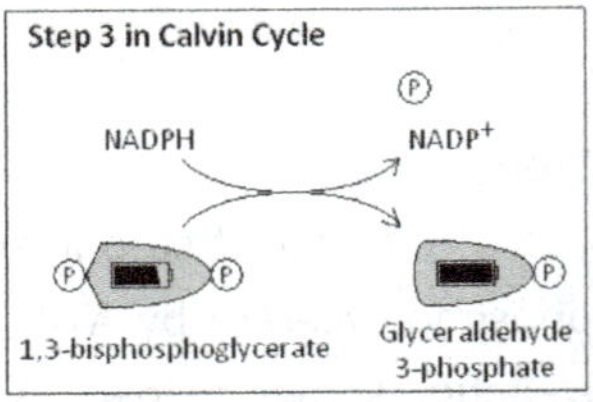

By this point, the glyceraldehyde 3-phosphate molecule that was created has been the recipient of the energy of both ATP and NADPH. That makes glyceraldehyde 3-phosphate *very* energetic. The plant takes two of these little balls of energy and makes sugar with them.

Two Glyceraldehyde 3-phosphate molecules → Glucose (the plant's energy for a rainy day)

Our discussion of the Calvin Cycle is not quite over though. At the beginning of the cycle, we started with ribulose bisphosphate, our "RuBi" molecule. The plant has to recreate this molecule somehow, or it can't make any more of these reactions work. So, let's see how the plant recreates RuBi.

Step 4:

As it turns out, not all of the very energetic glyceraldehyde 3-phosphate molecules are used to make sugar. Each time a bunch get made, the chloroplast takes 5 of them and mixes them together in a complex set of reactions to recreate 3 RuBi molecules. Three more ATP molecules are required to do this. The next image is what it looks like:

Step 4 in Calvin Cycle (many complex reactions)

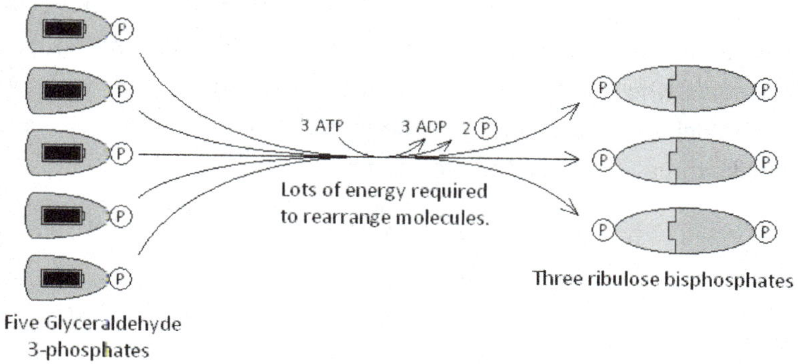

It is natural at this point for some confusion to creep into the situation, usually because the complex organic chemical structures complicate things. Also, you might be thinking something like: if it takes five glyceraldehyde 3-phosphates to make only three RuBi molecules, aren't we rapidly running out of molecules?

To make sure we clear up possible confusion, let's switch entirely to the little cartoonish symbols and go back over the reactions and keep track of the number of molecules. If you look at Step 1, each RuBi makes two smaller molecules. If you follow these molecules through Steps 2 and 3, each one becomes one glyceraldehyde 3-phosphate:

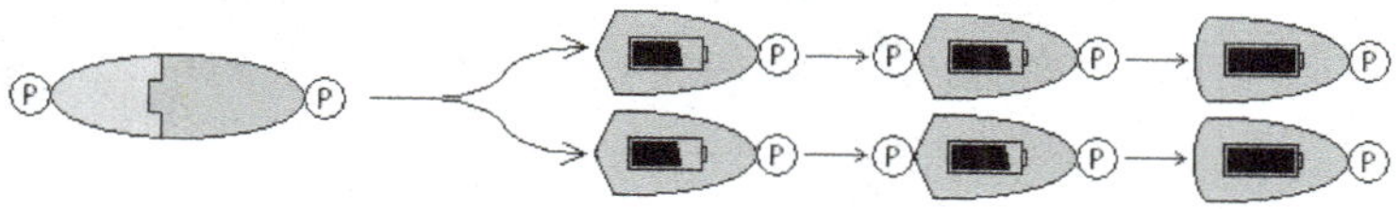

Or, in summary:

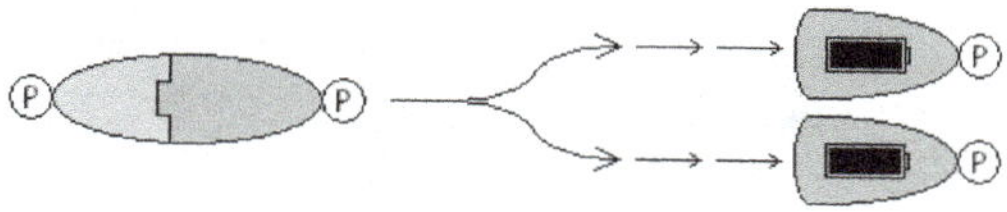

As this last image shows, each starting RuBi makes two glyceraldehyde 3-phosphates. Now imagine that we start the cycle with *three* RuBi molecules. Since each one ultimately makes two glyceraldehyde 3-phosphate molecules, we would end with six glyceraldehyde 3-phosphates at the end. We then can take five and remake three RuBi to do the cycle again.

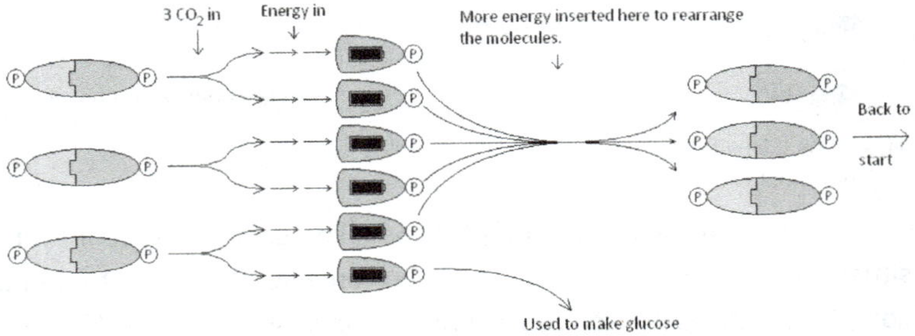

As this last image shows, the Calvin Cycle starts with three RuBi, absorbs three CO_2 as well as a bunch of energy, and creates one extra glyceraldehyde 3-phosphate that is used to make sugar. The rest gets turned back into three molecules of RuBi to start the cycle again. Here is a summary of the Calvin Cycle:

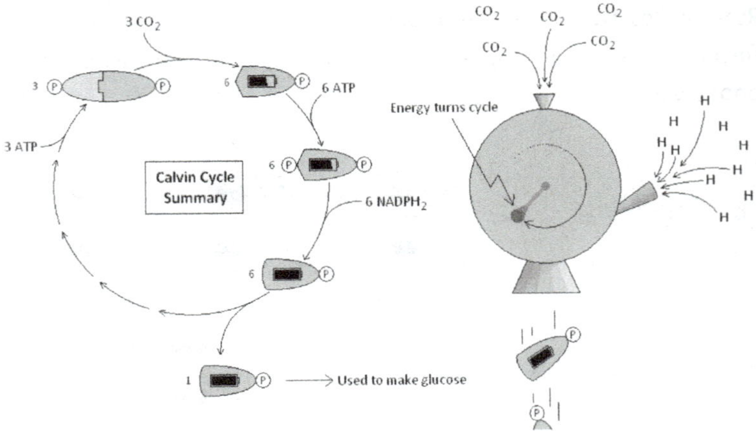

So to summarize our entire discussion of photosynthesis: the energy from sunlight is used to take CO_2 and water, and combines them together to make O_2 and glyceraldehyde 3-phosphate. This last molecule is used to make glucose. Here is the overall summary:

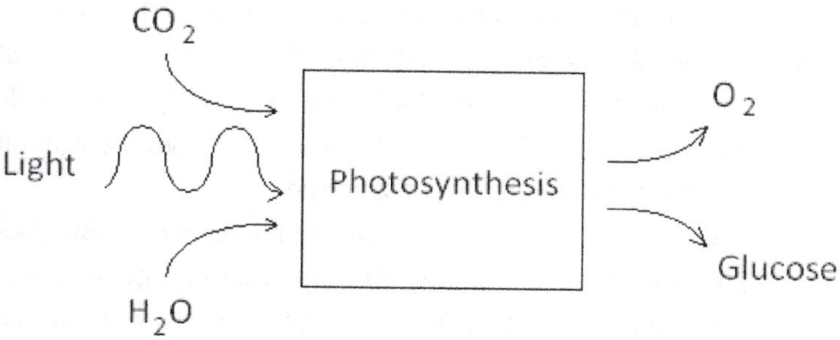

From this image, we see that CO_2 and water from our atmosphere is converted to O_2 and glucose. This simple innocent appearing little process is the reason that our planet has so much more oxygen and so much *less* CO_2 than other planets. It's how almost all life on Earth gets its energy.

In the next few units, we'll shift our attention to the chemistry that lets plants and animals pull the energy back out of glucose.

Summary

In this section, we talked about energy and photosynthesis.

- In casual terms, energy is the ability of one thing to change another. There are many forms of energy, and almost all are relevant to the processes of life at one point or another.
- Nearly all the energy that fuels life on Earth starts with sunlight. "Light" is part of a spectrum of energy waves called electromagnetic waves, or EM waves. The visible light that we see is only a part of that spectrum.
- The sun's visible light is captured by plants in photosynthesis. Plants convert this energy into chemical energy as glucose. They consume glucose for energy whenever there isn't enough light to get energy.
- Plants do not absorb green light. Green light bounces off and gives plants their color.
- Photosynthesis is divided into two sets of reactions: the light reactions and the dark reactions. The light reactions require the sun's light. The dark reactions take the energy that was captured during the light reactions and converts it into chemical form. Another name for the dark reactions is the Calvin Cycle. These reactions don't necessarily take place in the dark, but are named that way because they don't require light.
- The light reactions begin when a chlorophyll molecule is hit by incoming light. Chlorophyll is found inside a protein complex called a light harvesting complex, or LHC. These are found packed together with other LHCs in the thylakoid membranes of chloroplasts. The entire collection of LHCs is called a photosynthetic unit, which has a special "reaction center." When the light hits chlorophyll, one of its electrons gets excited and bounces through the complex of LHCs until it

hits the reaction center. Once there, the electron moves through other membrane proteins and finally arrives at an empty $NADP^+$ molecule. When the molecule captures two such electrons, it becomes NADPH and is used to fuel the Calvin Cycle.

- Each LHC that donates an excited electron to a $NADP^+$ molecule now has an electron deficit. This is filled by an analogous process taking place in a nearby similar set of LHCs. These LHCs now have an electron deficit. To solve the problem, this second set of LHCs steal electrons from water, and the oxygen atom that results pairs up with other oxygen atoms to make O_2 gas. Since plants are not gas tight, some of the O_2 is always leaking out into the environment.
- In addition to the energetic NADPH molecules that are made, the process of the electron bouncing through the membrane proteins causes a H^+ gradient to be produced inside the thylakoid interior. These crowded H^+ ions move through a channel in the membrane in order to escape back out, and in the process they provide energy needed to convert ADP into ATP.
- In the Calvin Cycle, CO_2 is pulled from the air and combined with ribulose bisphosphate. In the series of reactions that result, six molecules of glyceraldehyde 3-phosphate are produced. Five are recombined to recreate the starting ribulose bisphosphate molecules. One of the glyceraldehyde 3-phosphates is set aside and combined with another glyceraldehyde 3-phosphate (created during the next cycle) to make glucose, which is stored for later use when energy is needed. The NADPH and ATP generated in the light reactions are used to fuel the Calvin Cycle.

Unit 8

Oxidation and Reduction

In this section we cover:

1) Oxidation and Reduction
2) Oxidation glucose
3) Overview of ATP, its uses, and how it's made

In the last unit, we talked about energy and how it gets stored in glucose during photosynthesis. In the chapters following this unit, we will be talking about how energy is released from glucose. Until now, however, we haven't really commented much on why one molecule has more energy than another. We'll need to address this if we are going to have a deeper understanding of life and its energy patterns. The discussion centers on oxygen, which has a vital role in the movement of energy in living material. Oxygen is constantly jockeying for electrons in a process called **oxidation** and **reduction**, and this competition fuels practically all living reactions around you.

The oxygen molecule, O_2, or O=O, is a very unstable molecule. Oxygen likes to hog electrons, but if it's attached to *another* oxygen, neither win in the tug-of-war for electrons. This is a downright unhappy situation, leading to a very short molecular marriage. The O_2 breaks down whenever given the chance, exchanging partners for another more generous atom as soon as possible, and this releases the "tension" as energy. The fact that O_2 is so unstable is why you don't see much O_2 on other planets, like Venus and Mars. Here on Earth, we have plants that produce it during photosynthesis, so there is a large supply of unhappy O_2 in the air. This miserable oxygen is constantly butting into the business of other molecules, eye-balling neighboring chemicals for a chance to jump in to bonds and hog electrons. For example, when oxygen sees a carbon bonded to hydrogen, it sees an opportunity to be much happier. . .

Oxygen is *much* happier when it mixes with carbon and hydrogen, because it is easy to pull electrons from them. The more oxygen gets attached to C and H, the more energy gets released. In fact, that is precisely what happens when things burn. Oxygen gets into bonds and hogs all the electrons, releasing a LOT of energy in the form of heat and light.

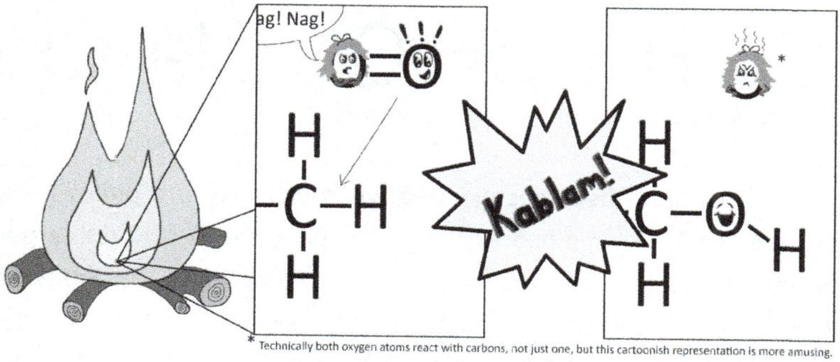

* Technically both oxygen atoms react with carbons, not just one, but this cartoonish representation is more amusing.

We say that the more oxygen that is attached to a carbon, the more **oxidized** the carbon is. The more oxidized it is, the more energy has been released. Conversely, this also means that the more oxidized a carbon is the less energy it has left to release. In other words, carbon molecules that have few oxygen atoms have high potential energy. Carbon molecules that have a lot of oxygen have low potential energy.

For example, here is what it would look like when butane oxidizes:

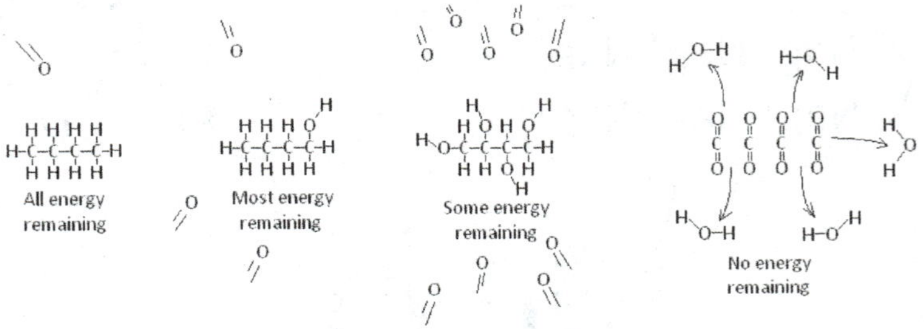

The oxygen keeps invading until the butane becomes so saturated with oxygen that is it broken into four CO_2 and five H_2O. At this point, butane has become fully oxidized and can release no more energy. The word "oxidized" is actually an unfortunate word, because oxygen isn't the only molecule that can cause this to happen. More globally, the word "oxidized" means that the atom has an electron pulled away from it, resulting in a positive or partial positive charge.

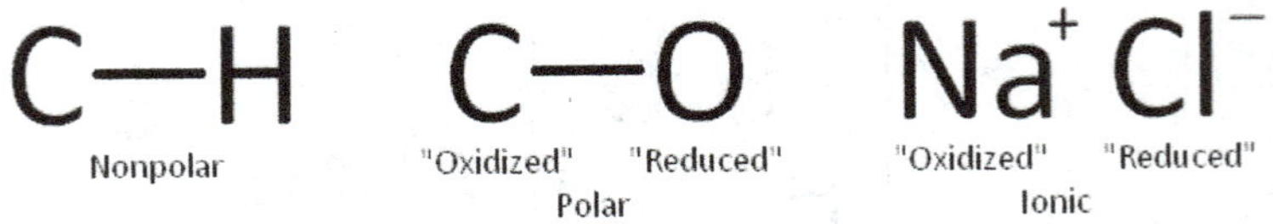

An atom becomes *oxidized when electrons are pulled away from it.* That is at the heart of the concept. Oxygen oxidizes carbon because it hogs electrons away from it. So, what do we say about the atoms that pull electrons toward them? We say that they become **reduced**. These atoms have electrons pulled toward them and have a negative charge or partial negative charge.

In summary:
Gain an electron ----> "Reduction"
Lose an electron ----> "Oxidation"

Ok, that said, let's look at glucose. It has six carbons. To remove as much chemical energy as possible, we have to oxidize it to the point where it is broken down into individual CO_2 and water molecules, much like what happened to butane. A biological cell could, in theory, just throw the glucose into a fire and let O_2 rip it to shreds, releasing all that energy, but the energy released would be uncontrolled (and dangerous), increasing the temperature and boiling the cell. That strategy won't work! The cell needs to oxidize the glucose one step at a time, carefully allowing the energy to be released in other forms. That way, much less energy gets lost as uncontrolled heat. How should the cell store these bits of carefully released energy? The answer is simple. It uses a number of other molecules that act as "middlemen" in the process. These middlemen take the electrons one at a time *from* glucose and give them *to* oxygen. This allows energy to be released more slowly and carefully. During this highly controlled process, the middlemen take these little bits of energy and store them in a much more useful and practical form: ATP.

ATP is the short-term, day-to-day energy currency of the cell. In fact, if glucose were a gold brick, ATP would be a $10 bill and much more practical for pushing around reactions or moving materials. No one pays for things with gold bricks after all. We might store our wealth in gold, but we pay for things with smaller value denominations.

Super, but what makes ATP capable of being an energy molecule? Let's look at it for a moment and see.

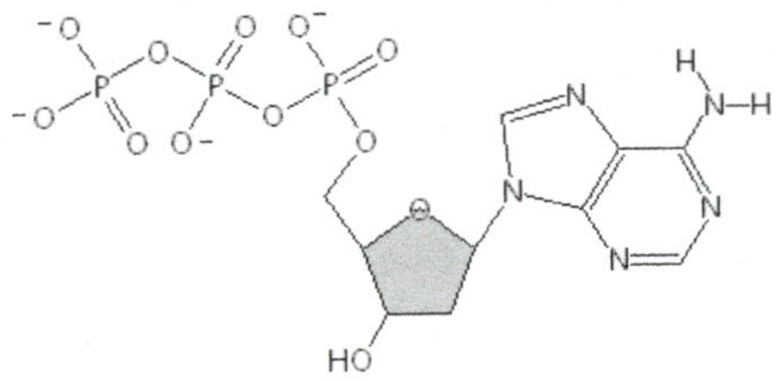

Adenosine triphosphate, "ATP"

Notice that ATP has three phosphate groups in it, strung up in a line. These phosphate groups are not happy in this situation, because each one is negatively charged. Negative charges repel, so each phosphate begrudgingly bonds to another, but you can bet they are counting down the time for when they can break up and leave. As such, ATP is lively and energetic. The cell makes ATP by taking the energy that is released when glucose is oxidized to force a phosphate group to add to the end of ADP in the following reaction:

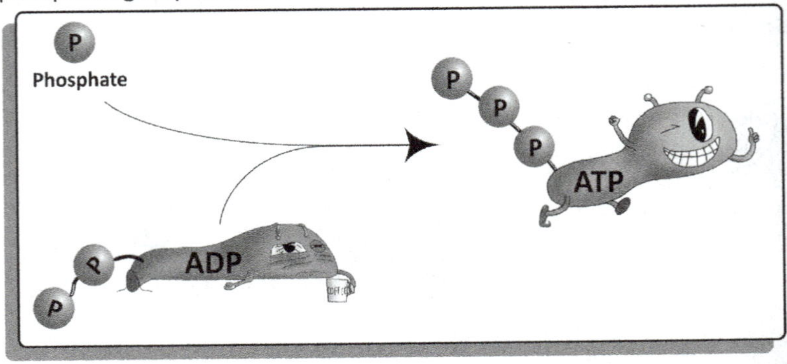

There are other forms of short-term energy currency in the cell, however. For example, energy can be stored as GTP (molecules similar to ATP), CTP (again similar), NADH, and $FADH_2$ molecules[21]. These other forms of short-term energy currency are used for somewhat specific purposes, whereas ATP is basically all-purpose. This helps a cell divide up the kinds of energy that it stores for short-term use. We'll discuss these molecules as they appear again later.

At any rate, the punch-line for this discussion goes as follows: Glucose represents the storage of lots of energy (much more than is useful to be released at once). So, the cell oxidizes it slowly, and the energy that is released gets stored in short-term currency, mostly as ATP. ATP has an amount of energy that is much more practical for the cell to use on a regular basis. ATP is the all-purpose energy currency. GTP, CTP, and other forms of short-term energy molecules are used for more specific purposes. Cells only oxidize as much glucose as they need to keep a regular supply of short-term energy currency around. The rest of the glucose gets set aside for later use.

Now that we have the basic process down, we'll spend the last two units seeing the details. . .

---

[21] NAD^+ = nicotinamide adenine dinucleotide, related to NADP from photosynthesis. FAD = flavine adenine dinucleotide. NADH and $FADH_2$ are their reduced forms.

Summary

In this section, we talked about oxidation, reduction and an overview of ATP.

- Oxidation occurs when electrons are pulled from an atom. These atoms become positively or partial positively charged.
- Reduction occurs when electrons are added to an atom. These atoms become negatively or partial negatively charged.
- Oxygen is commonly the atom that oxidizes organic molecules, though any molecule that is more electronegative than carbon may also oxidize organic molecules.
- When oxygen oxidizes carbon and hydrogen, energy is released. Oxygen likes electrons and releases energy when it gets them.
- When oxygen fully oxidizes carbon and hydrogen, CO_2, H_2O and all the energy is released.
- Fire occurs when O_2 oxidizes organic material in an uncontrolled fashion, releasing all the energy as heat and light.
- Living systems control the oxidation process by putting a middleman between the organic molecule and oxygen. The middleman takes the electron from the organic molecule and gives it to oxygen. This process releases energy in a controlled manner and stores it as ATP.
- ATP is the basic all-purpose energy currency of the cell. It is a more practical form of energy than glucose. Glucose is a storage form of energy, whereas ATP is the active form of energy that the cell uses for day-to-day chemical reactions.
- The cell uses a few other energy molecules as well, such as GTP, CTP, NADH, and $FADH_2$. These molecules have limited use, however, and are restricted to only a few processes in the cell.

Unit 9

Respiration, Part 1: Glycolysis

In this section we cover:

1) Glycolysis

2) NAD^+/NADH balance, and lactic acid production

Having seen photosynthesis and the processes of oxidation and reduction, we are now in a position to talk about how the biological world (including plants) breaks down glucose, a complex set of reactions known as **respiration**. In this process, we do the reverse of the overall reaction that went into making the glucose, namely:

$$C_6H_{12}O_6 + 6O_2 \text{ ----> } 6CO_2 + 6H_2O + \text{energy}$$

This happens in two sets of reactions. The first is a more primitive set of reactions known as **glycolysis**. In glycolysis, a small amount of the energy and a little water is released from glucose, transforming the glucose into **pyruvic acid** (also called **pyruvate**). The second set of reactions is a cycle that breaks the pyruvate down into CO_2, more water, and a LOT more energy. This cycle of reactions has many names: Krebs cycle, the citric acid cycle, and the tricarboxylic acid cycle. We'll refer to it as the citric acid cycle.

Here is an overview of respiration:

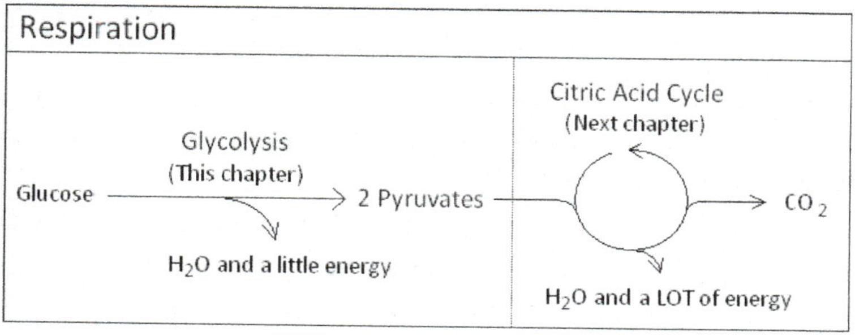

If you look at the overall reaction, glucose is broken down into CO_2, water, and energy, just like the butane was in the last chapter. The net effect is the same as if O_2 just went in and ripped glucose apart, but as we noted in the last chapter, the cell has to be careful to release the energy in an organized fashion. If the cell lets O_2 have free reign to mingle with the carbon and hydrogen, too much energy will be released too quickly and in a disorganized fashion. In that sense, glycolysis and the citric acid cycle are the mechanisms the cell uses to keep this energy harvesting under control. We will focus on glycolysis in this unit, and then look at the citric acid cycle in the next.

Glycolysis

Glycolysis means roughly "splitting glucose." During the course of glucose oxidation, one glucose gets broken in half, and the two parts are then converted into two pyruvate molecules. During the course of these reactions, a small amount of energy is harvested by the cell. In glycolysis, O_2 is not used to complete the oxidation. Rather, a molecule called NAD acts as the middleman, taking electrons from glucose and transporting it to where oxygen is. *That* story picks up in the citric acid cycle, which we'll discuss later. In addition to the NAD, a net total of two ATP are produced, but to understand how this happens, we'll need to examine the details.

There are ten reactions in the glycolytic pathway. The first three reactions basically prepare the molecule for being split in half, which happens in the fourth reaction. The remaining reactions pull a small amount of energy from the pieces and spit out two pyruvate molecules at the very end.

Let's examine these ten steps. The following images are an overview of the reactions, each step labeled numerically, 1 through 10. Following these images, a summary is displayed as well.

Step 1:

In the first reaction, the enzyme **hexokinase** takes one ATP and uses it to attach a phosphate onto glucose, making **glucose-6-phosphate**. The ATP is converted into ADP and floats away. Remember, the reaction ATP --> ADP releases energy, and in this step the energy is invested in the glucose molecule when the phosphate is added.

You may be wondering what the point of this step is, considering it *spends* ATP rather than making it. This is a small investment of energy (and not the only one) that the cell will eventually make back before glycolysis is over.

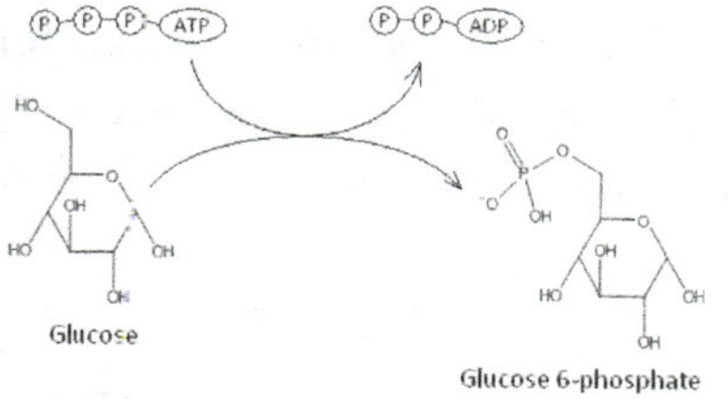

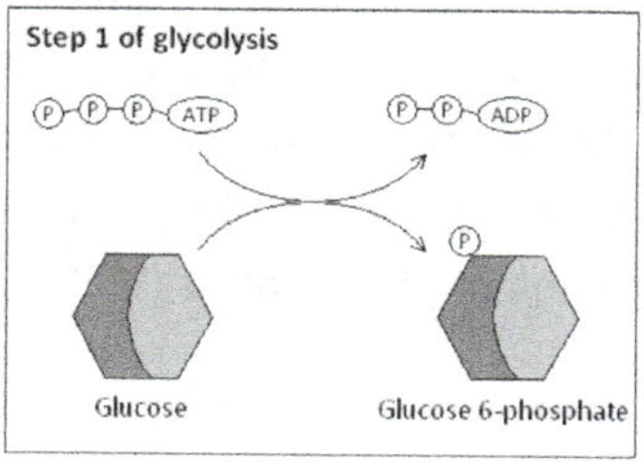

Step 2:

In the second step, the glucose-6-phosphate is rearranged internally into **fructose-6-phosphate** by an enzyme called **phosphoglucoisomerase.**

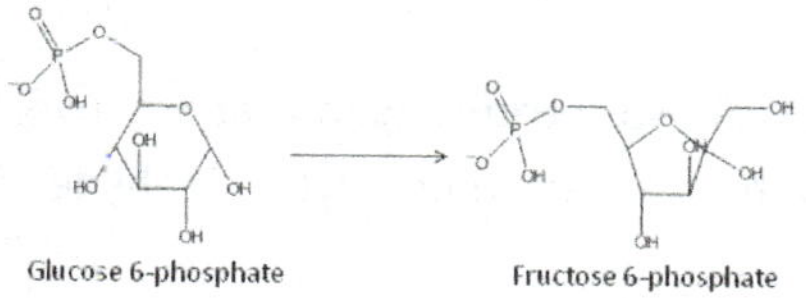

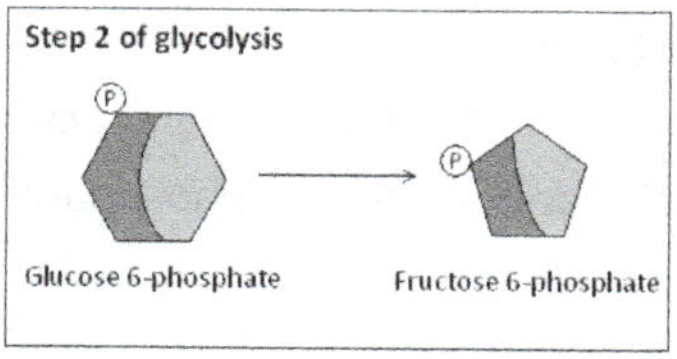

Step 3:

Next, we take this rearranged molecule and stick another phosphate on it. The enzyme that does this is **phosphofructokinase**. The molecule that results is **1,6-bisphosphate**.

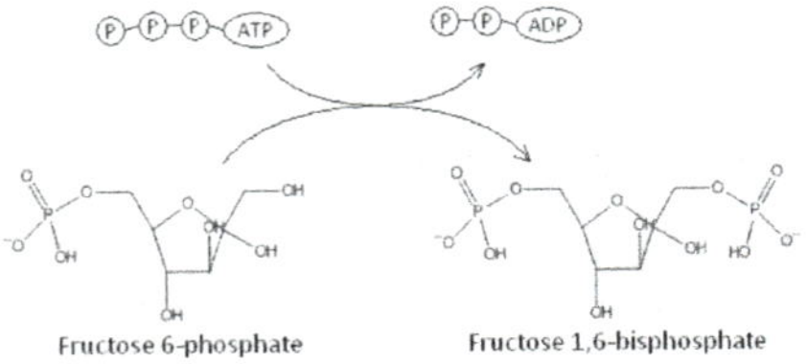

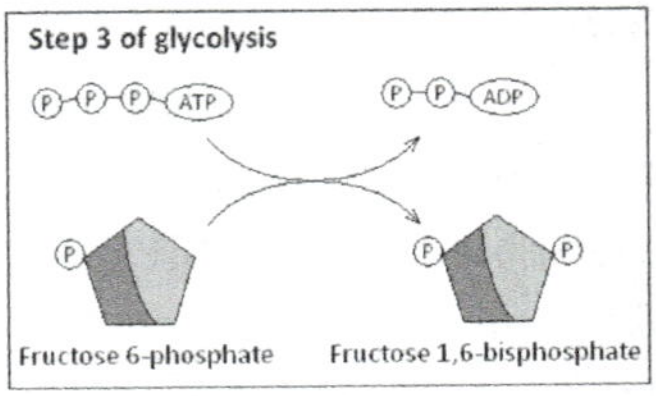

Again the cell invests another ATP at this point. So far, the cell is two ATP in debt, but it won't stay that way.

Step 4:

In reaction 4, we finally break the fructose 6-phosphate molecule in half using an enzyme called **aldolase**. This creates two molecules, each with a phosphate on the end. The two molecules are not identical, however. The first is called **dihydroxyacetone phosphate**, and the second is called **glyceraldehyde 3-phosphate**. The glyceraldehyde 3-phosphate is the same molecule that is produced during photosynthesis (unit 7).

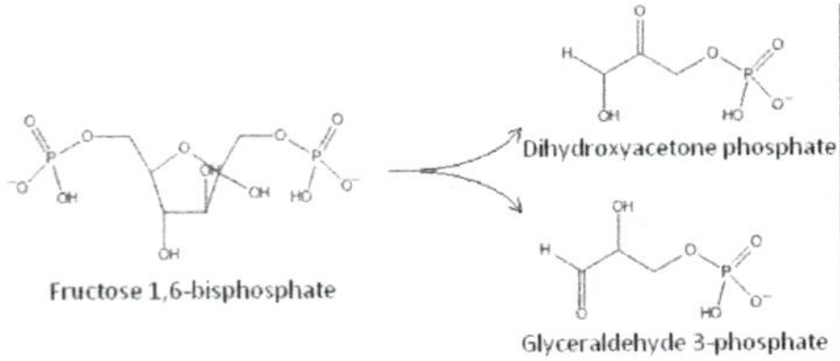

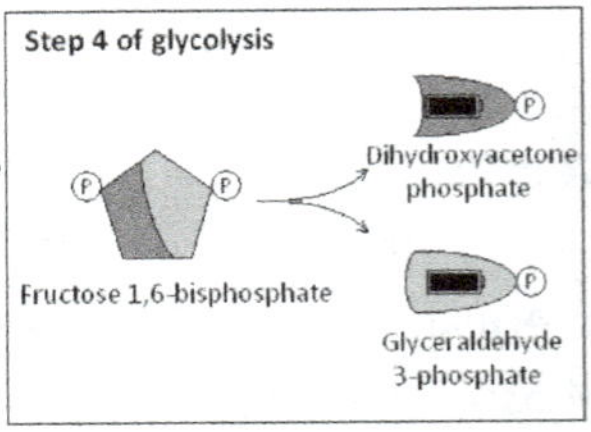

Step 5:

Next an enzyme called **phosphotriose isomerase** converts dihydroxyacetone phosphate into yet another glyceraldehyde 3-phosphate.

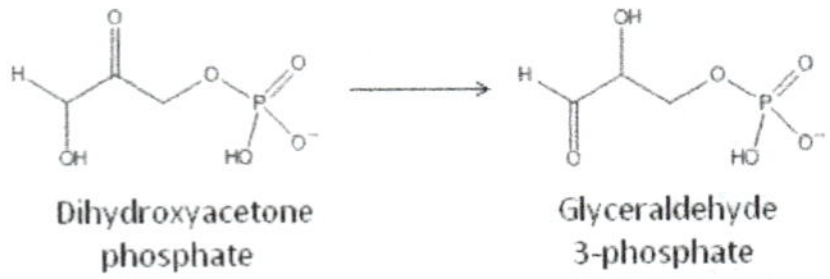

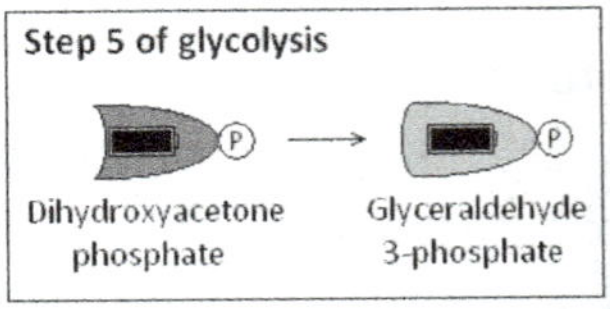

At this point, it is important to understand that we now have *two* copies of glyceraldehyde 3-phosphate: one was created when the fructose 6-phosphate was split in half, and the other was created by converting the dihydroxyacetone phosphate molecule. This means that all the following reactions happen *twice* per glucose molecule that originally enters into glycolysis.

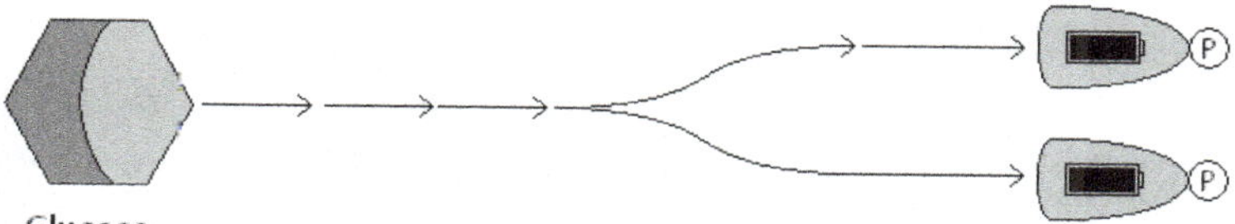

If you'll recall from the discussion of photosynthesis, plants put two glyceraldehyde 3-phosphates together to make glucose in an effort to *store* energy. Considering that, it shouldn't be surprising that we can get two of these molecules back out again by breaking down the glucose to *release* energy for use.

Step 6:

Here is where the action begins, so hold on to your hats! In the sixth step each glyceraldehyde 3-phosphate is oxidized. From our discussion on oxidation, this means an electron is pulled from the molecule. This electron is given to a molecule called NAD^+, a relative of the $NADP^+$ we encountered when we studied photosynthesis. By receiving the electron in this step, the NAD^+ becomes reduced to NADH and is more energetic. This is the first energetic molecule that emerges from the glycolysis reactions. It looks like our initial investment of two ATPs is starting to pay off. Anyway, the reaction also adds another phosphate to the glyceraldehyde 3-phosphate. The product that results is **1,3-bisphosphoglycerate.**

The enzyme that performs this step is called **glyceraldehyde 3-phosphate dehydrogenase**. The word dehydrogenase refers to an enzyme that participates in a reaction that pulls hydrogen from a molecule. If you look at the image that follows, a phosphate is added, and an H^+ is freed. The release of that H^+ is what the name "dehydrogenase" indicates.

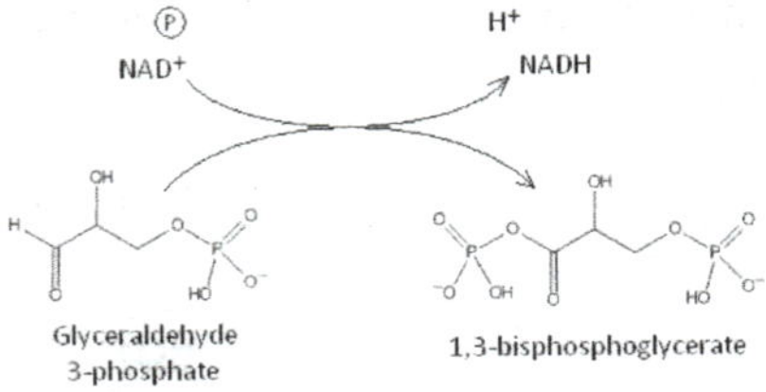

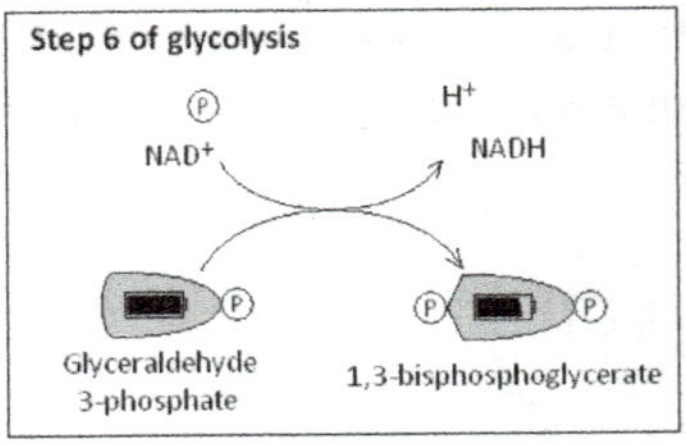

Step 7:

We are now ready to generate the first ATP! The 1,3-bisphosphoglycerate is an unhappy molecule with two phosphates stuck on it. It has enough energy that it gives one of these phosphates to ADP, creating ATP. The molecule that forms is **3-phosphoglycerate.** Remember that Steps 6 and 7 happen twice for every glucose we use, because two glyceraldehyde 3-phosphates were made from the original glucose. That means we've made back our initial investment of ATP. So far, we've netted two NADH energy molecules. Yay!

The enzyme that performs this step is called **phosphoglycerate kinase**. I figured I would tell you that so you don't lay at night wondering about it.

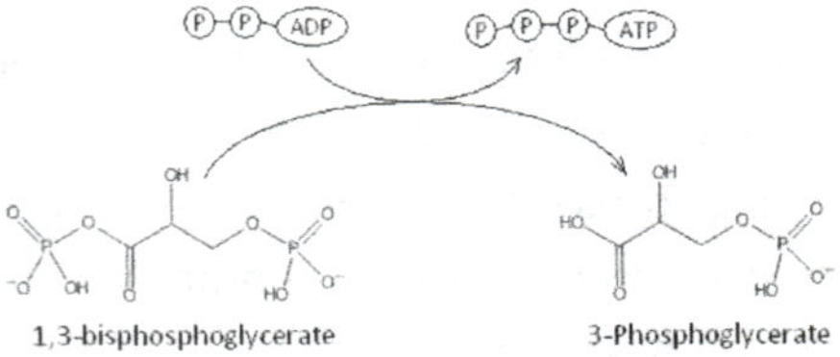

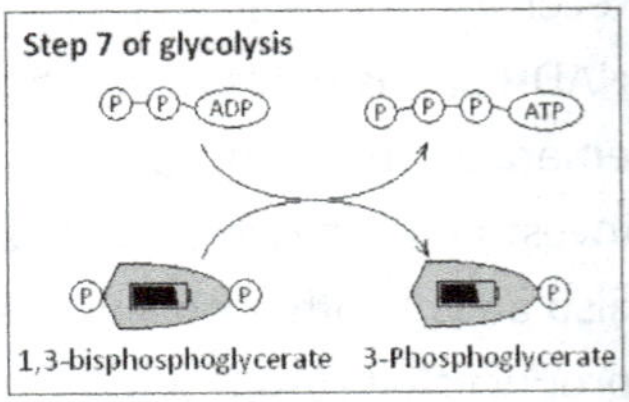

Step 8:

Now we do an internal rearrangement using the enzyme **phosphoglyceromutase**. This changes the 3-phosphoglycerate into **2-phosphoglycerate**.

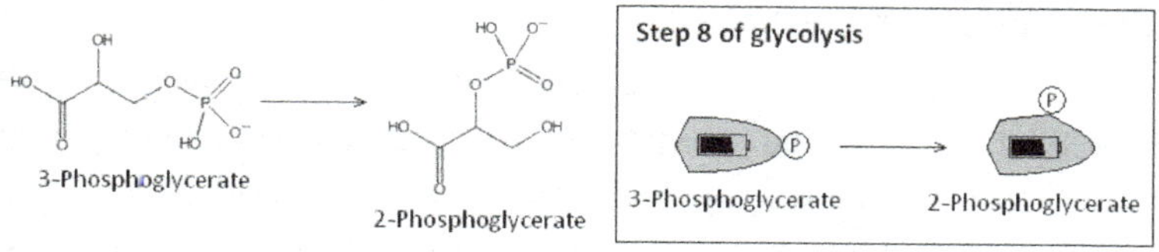

Step 9:

Then, we do yet another rearrangement. The enzyme **enolase** performs this task. Notice that a water molecule is produced here. This is our first evidence of the overall oxidation end products: CO_2 and water.

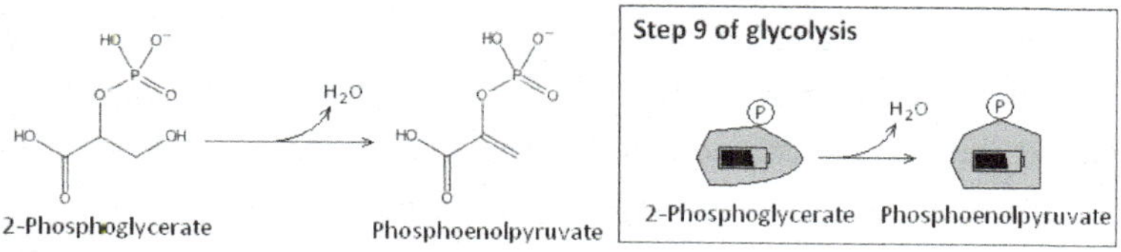

Step 10:

Now, after all that mess, we are finally ready to reap the rewards of glycolysis. The enzyme **pyruvate kinase** pulls the last of the phosphates from the **phosphoenolpyruvate** to turn ADP into ATP. Remember, this happens twice per glucose, so this step yields another two ATP, one for each phosphoenolpyruvate. Our initial investment of two ATP was already paid back, so these new two ATP are pure profit! Woohoo! The final molecule that we get is pyruvate.

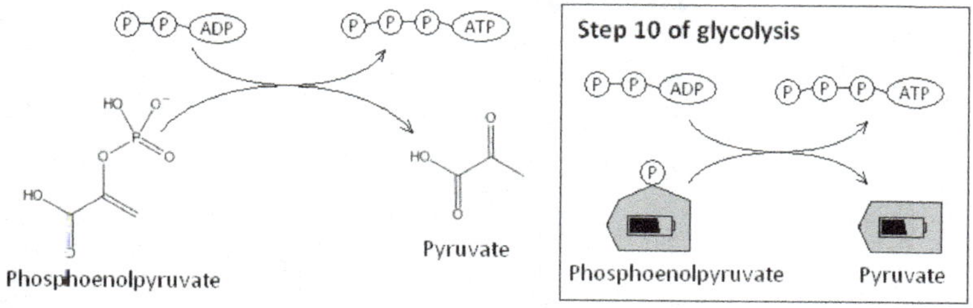

Glycolysis summary:

Let's put all these reactions together into one picture to see how they work. If you examine the image below, you'll see glucose (the first hexagon shape) going through various changes. These changes ultimately require the input of two ATP in order to get the reactions started. So, that puts the living system in "energy debt" with negative two ATP in its account, so to speak. After the glucose gets broken into the smaller pieces, each individual piece eventually releases two ATP and one NADH, earning the living system four ATP *back*, as well as two molecules of NADH. To summarize, the living system spends two ATP and gets four back, so it earns two overall. The smaller picture in the bottom right corner of the following image shows the net effect: one glucose, two NAD, and two ADP enter into glycolysis, yielding two pyruvate, two ATP, and two NADH.

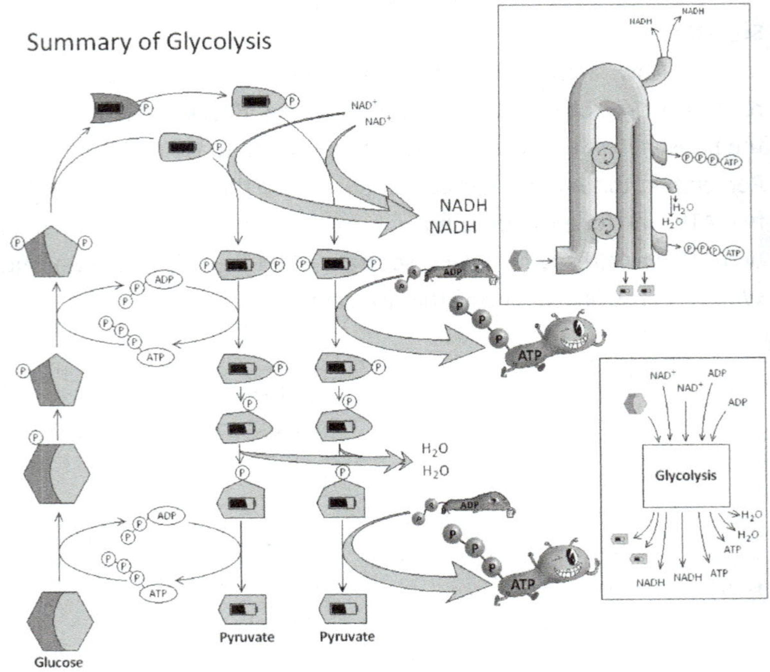

The story of how living systems get energy out of glucose is not over yet though. The two pyruvate molecules that result from glycolysis still have a lot of energy left in them! From here, the two pyruvates and the two NADH molecules move on to the next set of reactions, the citric acid cycle, where the rest of the energy will be released.

NAD/NADH Balance and Lactic Acid

There is, however, one last thing we need to touch upon before we conclude the unit. During the course of glycolysis, the living system converts NAD^+ into NADH. The NADH molecule is very energetic, and when it is used in other reactions, it gives up its energy and is converted back into NAD^+. It then returns to glycolysis to get charged up again. Unfortunately, although NADH is energetic, it is not a very convenient form of energy for the living system to use, not like ATP is anyhow. If ATP was a $10 dollar bill, NADH would be a special token that can only be used at three convenience stores across the country.

The point is: there are only a few reactions that use NADH as an energy source, and if these reactions do not need energy, the NADH would just sit there collecting dust and getting in the way. If they just sit there, it means they do not get converted back into NAD^+. Therein hides the real problem: the cell could run out of NAD^+! Glycolysis MUST have NAD^+ or it can't proceed, so the cell must always have a way to convert NADH back into NAD^+ or else it can't get any more energy out of glucose. If glycolysis stops, the living system will die.

As it turns out, there are two standard ways in which NADH is converted back into NAD^+. The first (and more preferable way) occurs during the citric acid cycle, which requires the presence of O_2. The second way is for NADH to react with the newly made pyruvate, turning it into **lactic acid**. The reaction looks like this:

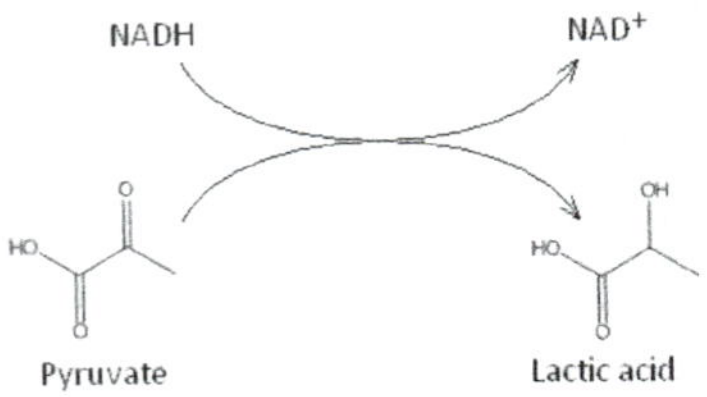

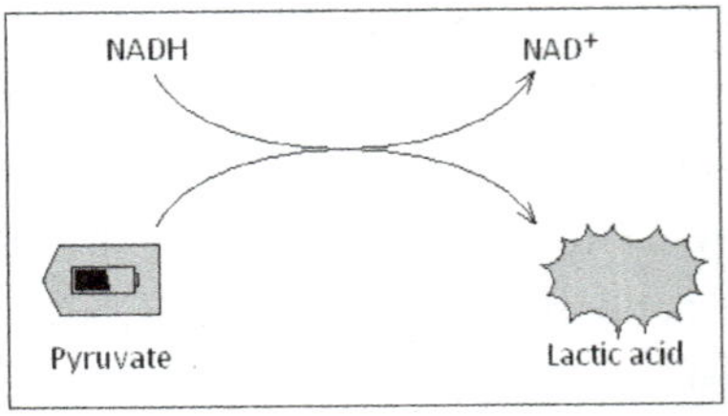

When O_2 is present, both pyruvate and NADH will proceed together into the citric acid cycle, where a *great* deal of energy is released. When this is the case, respiration is called **aerobic respiration**. When O_2 is *not* present, pyruvate and NADH react together. When this happens, respiration is called **anaerobic respiration**, and it doesn't produce nearly as much energy.

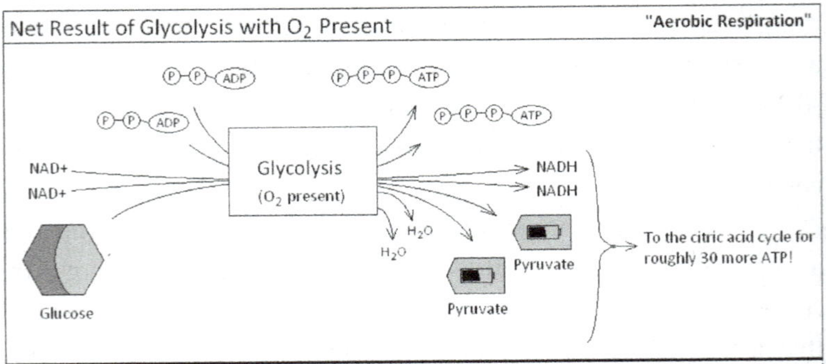

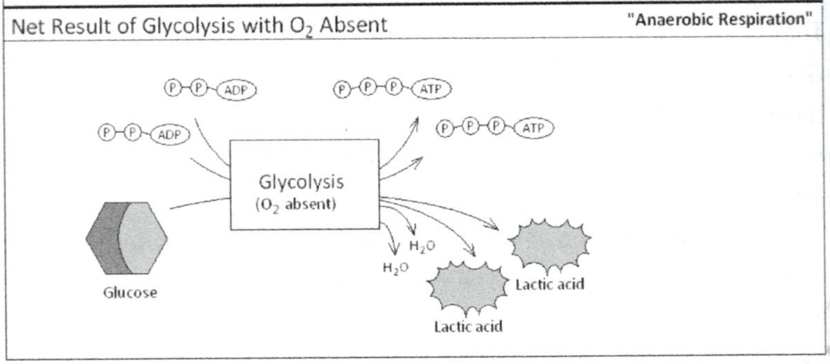

Due to the nature of anaerobic respiration, lactic acid forms any time the body is deprived of O_2, such as during suffocation, intense exercise, or in the moments following clinical death. When the body dies, for example, the heart stops beating and blood stops flowing. As long as it is flowing, blood delivers oxygen to the body. So, when the blood stops pumping, all the cells around the body begin suffocating. Since they can't get oxygen, they cannot do oxidative respiration and switch to reacting NADH and pyruvate together. This makes a lot of lactic acid. Since anaerobic respiration produces far less energy, some of the energy dependent organs, such as the brain, begin to rapidly shut down.

When you exercise vigorously, you begin using energy much faster than the blood can deliver oxygen to the muscles. In order to continue working, your muscles – now deprived of the necessary oxygen – switch to reacting NADH with pyruvate. This creates lactic acid and is why you get that familiar "burn" when exercising[22].

---

[22] The soreness you get days later is not due to lactic acid though. It is due to microscopic tears in the muscle that are healing.

Summary

In this unit, we talked about glycolysis and lactic acid production.

- Glycolysis means "breaking sugar," and it is the first part in a set of reactions called respiration. The second part of respiration is the citric acid cycle, discussed in the next unit.
- Glycolysis breaks glucose in half and partially oxidizes it in a controlled set of ten reactions. Two ATP, two NADH, two water molecules and two pyruvates are created.
- If oxygen is present, the pyruvate and NADH move on to the next set of reactions, the citric acid cycle, where much more energy is harvested. This version of respiration is called aerobic respiration.
- If oxygen is not present, the pyruvate is reacted with NADH to make lactic acid and NAD^+. This frees up NAD^+ to participate in another round of glycolysis. This version of respiration is called anaerobic respiration.

Unit 10

Respiration, Part 2: Citric Acid Cycle (a.k.a. Krebs Cycle or Tricarboxylic Acid Cycle)

In this section we cover:

1) The Citric Acid Cycle
2) The Electron Transport Chain
3) The total energy obtained from respiration

In the last unit, we talked about how glucose is transformed into two pyruvates during the ten reactions of glycolysis. We also touched on the issue of keeping NAD^+ and NADH in balance when there is no oxygen to be the acceptor of the electron that NADH carries around. Without oxygen, the pyruvate that is made is turned into lactic acid.

In this section we talk about what happens in the **citric acid cycle**, where the rest of the energy in pyruvate is converted into water and CO_2. Together, the citric acid cycle and glycolysis achieve the reverse of photosynthesis:

Photosynthesis

$$6CO_2 + 6H_2O + \text{energy (as light)} \text{ -----> } 6O_2 + C_6H_{12}O_6$$

Respiration (glycolysis + citric acid cycle)

$$6O_2 + C_6H_{12}O_6 \text{ ----> } 6CO_2 + 6H_2O + \text{energy (chemical)}$$

The summary of these two reactions is: 1) autotrophs do photosynthesis, storing the energy of light in glucose, and 2) both autotrophs and heterotrophs then take the glucose and break it down to get the energy back out.

Citric Acid Cycle

The Introductory Reaction

At the end of glycolysis, glucose has been cracked in half and turned into two pyruvate molecules. From here, these two pyruvates are taken into the mitochondria (see unit 6) where the citric acid cycle awaits them.

Introductory Step:

The first reaction to take place is the addition of a chemical side group called **Coenzyme A** (CoA) to pyruvate. CoA has a fairly complex structure and is related to nucleic acids, as shown to the right. CoA was introduced in Unit 11 of the first topic text. Its addition to pyruvate is catalyzed by the enzyme **pyruvate dehydrogenase**, as shown below:

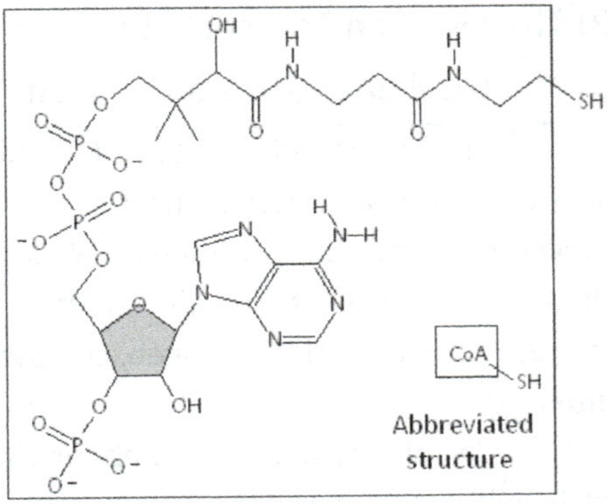

Coenzyme A

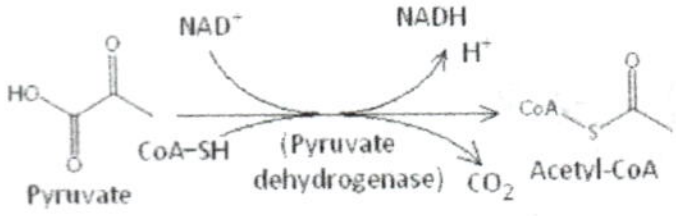

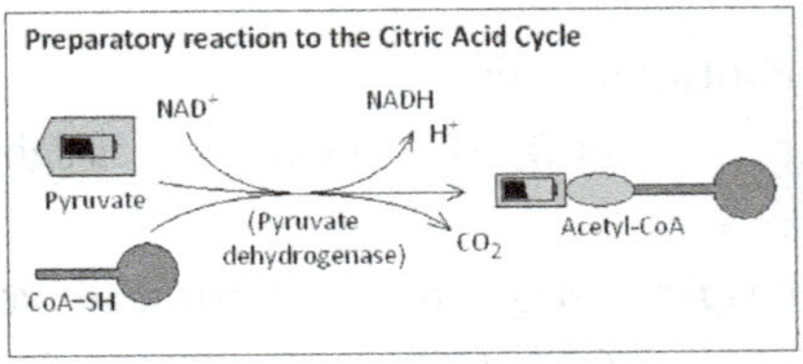

As a side note: pyruvate dehydrogenase is a heavy-hitter in courses like biochemistry and on standardized exams, especially in the medical profession. If that is your chosen goal[23], then you may as well get used to this enzyme.

---

[23] Oh you poor soul....

Anyway, as you'll notice from this reaction, one of the CO_2 in pyruvate gets set free. Since two pyruvates per glucose are created, two CO_2 total are set free at this stage. Remember that glucose started with six carbon atoms. This means that we've carried out complete oxidation (gotten the most energy possible) out of two of the six carbons in glucose. The citric acid cycle frees the remaining four.

Acetyl-CoA is the product of this initial step. For practical purposes, acetyl-CoA is the official start of the citric acid cycle. It is worth pointing out that acetyl-CoA is also produced during the breakdown of most lipids and many proteins as well. Basically, the living system recycles the energy it took to make the lipids and proteins[24] by breaking them down to make acetyl-CoA and/or pyruvate, which get funneled into the citric acid cycle and broken down into CO_2, H_2O and energy. In that sense, the citric acid cycle is not limited to the breakdown of glucose, but is a universal set of reactions that uses many of the cell's chemicals.

Common Route of Biomolecule Breakdown

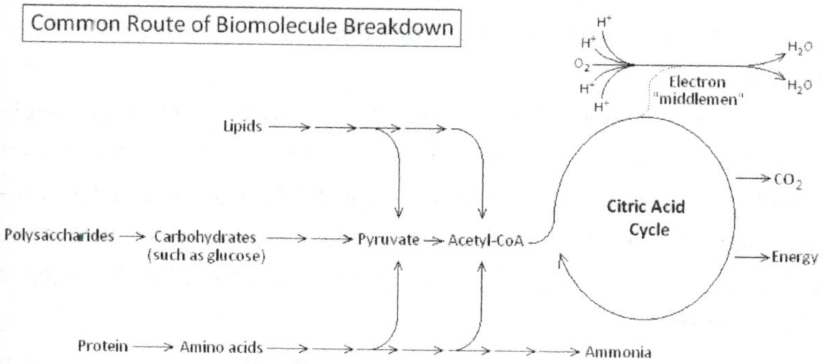

As this image shows, the citric acid cycle makes energy, CO_2 and electron "middlemen." These middlemen represent the majority

[24] Most of the amino acids can be broken into acetyl CoA, but not all.

of the actual energy to emerge from the citric acid cycle. These molecules carry electrons to oxygen, and this process unlocks a great deal of energy, as we'll see. Otherwise, the actual amount of energy that is produced from the citric acid cycle (labeled as just "Energy" in the picture) is very minimal.

You'll also notice from the image that when protein is broken down to enter into the citric acid cycle, ammonia is produced as a side product. The chemical formula for ammonia is NH_3, and this chemical gets made because of the nitrogen in amino acids. The citric acid cycle does not handle nitrogen compounds, so this chemical gets put off to the side to be removed by other means. In animals, such as humans, it is removed in the urine.

The Citric Acid Cycle Reactions

As with the Calvin Cycle and glycolysis, the individual reactions are listed here in sequence. Some introductory biology courses will not require you to memorize the reactions, but you may find this listing helpful if you have to study the reactions more in depth at a later time. Otherwise, a summary image is given at the bottom as well as a brief discussion of the overall effects of the citric acid cycle.

Step 1:

Once acetyl-CoA is formed, (regardless of which biomolecule it started from), it is now ready to enter into the citric acid cycle. In the first reaction, the acetyl part of the acetyl-CoA is added onto a pre-existing molecule called **oxaloacetate**. This addition forms **citric acid**, which gives the cycle its name. The enzyme that performs this step is **citrate synthase**.

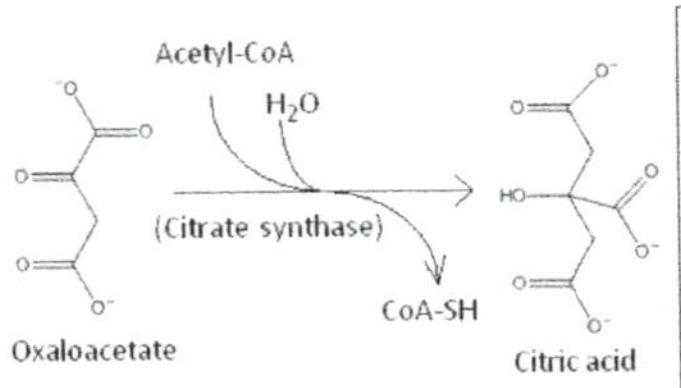

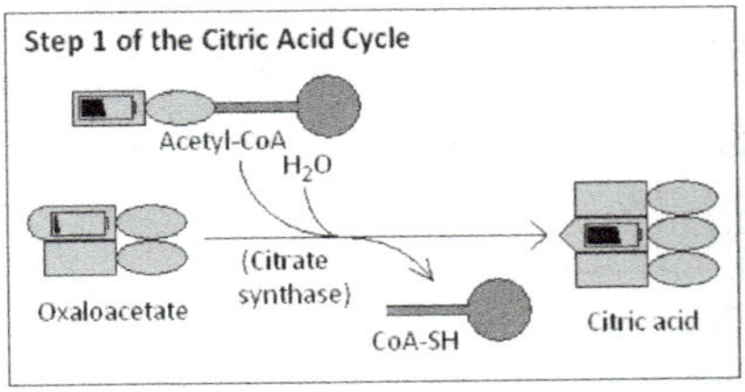

Step 2:

The next step is catalyzed by the enzyme **aconitase** and is a simple rearrangement of citric acid to **isocitric acid**. Apart from this minor rearrangement, nothing else of much excitement happens here.

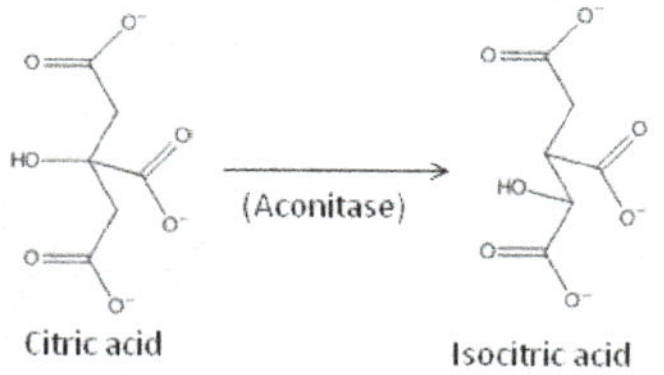

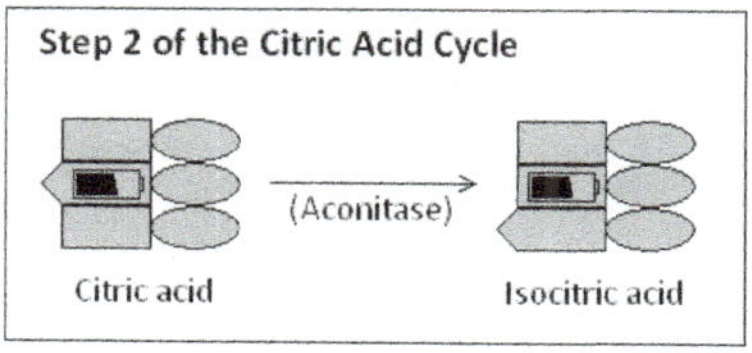

Step 3:

The next step in the cycle is performed by the enzyme **isocitrate dehydrogenase**. In this reaction, one CO_2 is released from isocitrate to make **α-ketoglutarate**. One NADH is created in the process. Remember that this happens once per pyruvate. Since there are two pyruvate made from each glucose, this step ultimately releases *two* CO_2 and *two* NADH molecules per glucose. This means that two of the final four carbon atoms are now accounted for. We only have two carbons left to release!

The NADH that is produced is one of the electron middlemen. It is highly energetic, and once made it floats off to accomplish its task of making energy by giving electrons to oxygen.

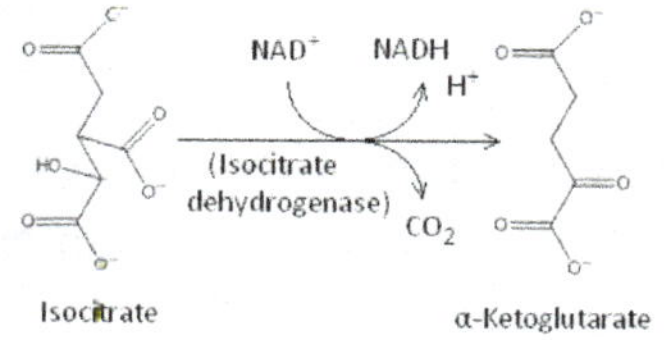

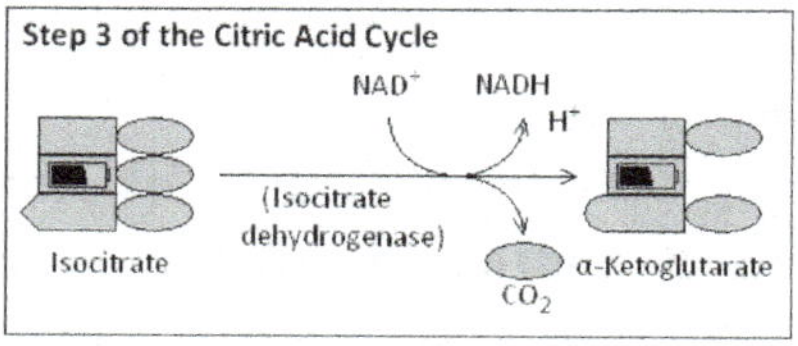

Step 4:

Step 4, catalyzed by **α-ketoglutarate dehydrogenase**, takes α-ketoglutarate, removes CO_2 and attaches CoA to make **succinyl-CoA**. Once again, this happens twice per glucose, so this accounts for the last two carbon atoms.

In addition to freeing the last carbon dioxide, this step also produces NADH, which floats off to make energy like the NADH in the previous reaction.

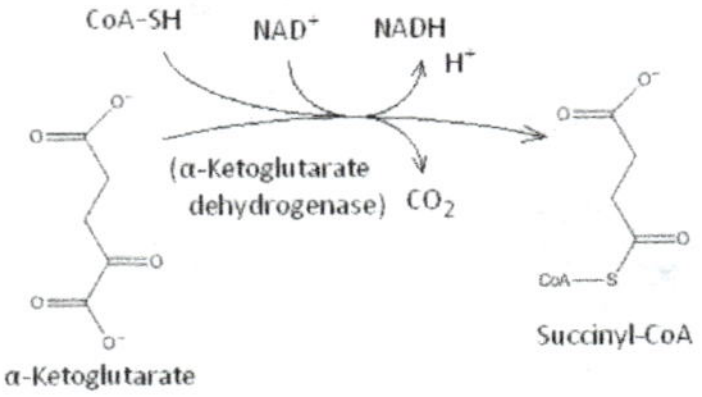

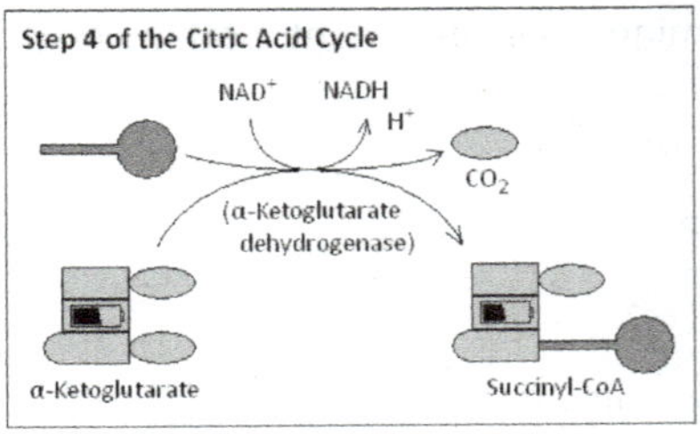

Since we've accounted for the last carbon, then it's entirely possible that the story of the citric acid cycle is over, and I've just filled the rest of this unit with random letters that just happen to make sentences. Well, for as nice as it would be for the story to be over, we still have a very big stumbling block to take care of: we need to regenerate oxaloacetate so we can run the cycle again! Oddly enough – as it turns out – the process of regenerating the oxaloacetate manages to squeeze a *large* part of the remaining energy left over from the pyruvate. It's weird – the carbons are gone, but the energy is still left lurking, like the ghost of glucose past. So let's keep going and see what kind of molecular brouhaha we can discover.

Step 5:

In step 5, **succinyl-CoA synthase** takes the succinyl-CoA that was created during the last reaction and pulls the CoA from the molecule to make **succinate**. This process releases a little energy, which the citric acid cycle uses to make GTP. If you'll recall, GTP is like ATP; it is an energy molecule, but with less universal use. Shortly after GTP is made, another reaction rapidly converts it into ATP.

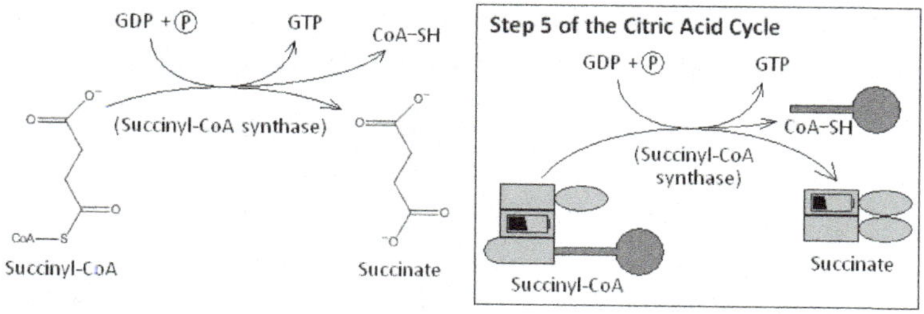

Step 6:

The next step to occur is the conversion of succinate to **fumarate**, which is catalyzed by the enzyme **succinate dehydrogenase**. This reaction creates one $FADH_2$ which is an electron middleman like NADH.

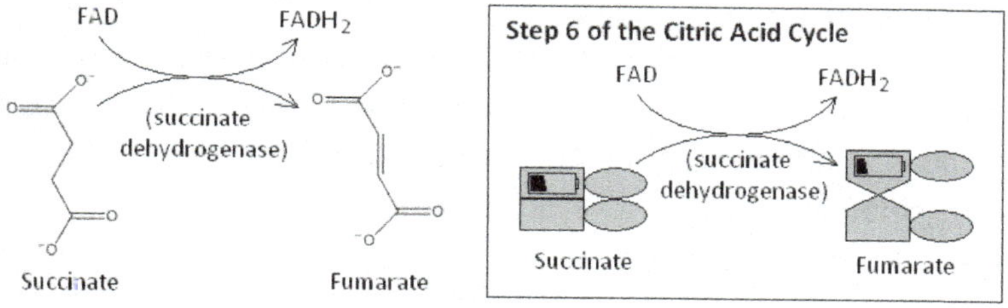

Step 7:

In step 7, the enzyme **fumarase** catalyzes the addition of water to fumarate to make malate. Apart from this, the reaction is otherwise unexciting.

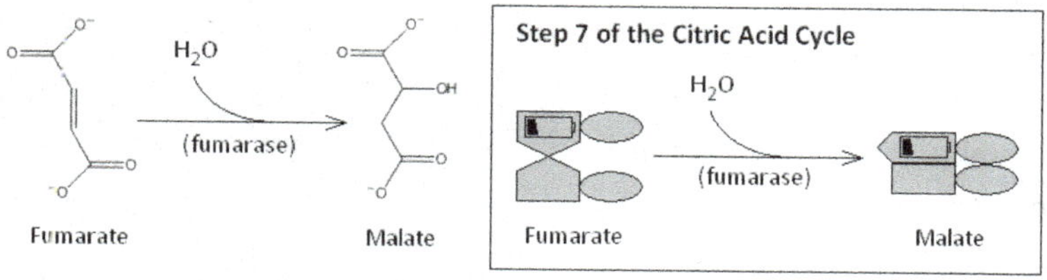

Step 8:

Step 8 is the last reaction in the cycle. This step makes the original oxaloacetate molecule (needed in step 1) so that the cycle can happen again. This reaction is catalyzed by **malate dehydrogenase,** which makes another molecule of NADH in the process as well.

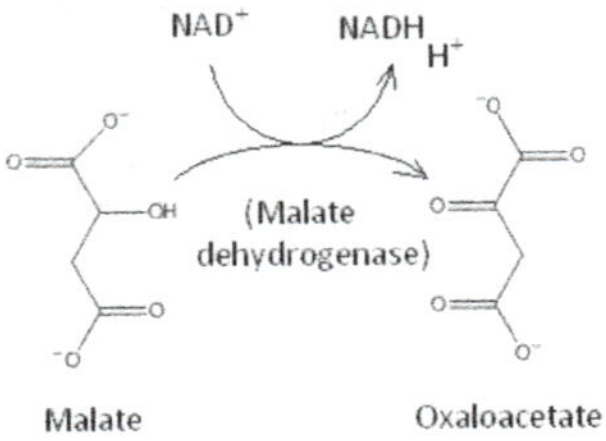

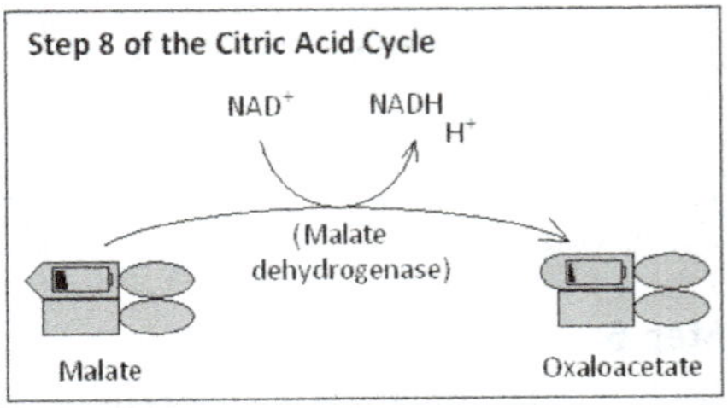

Not bad. Even though all the carbon atoms from the original pyruvate get released early in the cycle, we still pull 1 ATP, 1 $FADH_2$ and 1 NADH from the remaining reactions. Spooky.

Summary of the Citric Acid Cycle:

Let's take an overview of the reactions of the citric acid cycle. Pyruvate is initially converted to acetyl-CoA during the introductory reaction. This releases one CO_2. Within the next three steps of the cycle, another two CO_2 are freed. Two pyruvate are made per glucose, so ultimately six CO_2 are released during the introductory and first three steps of the citric acid cycle. That accounts for all six carbon molecules in the starting glucose.

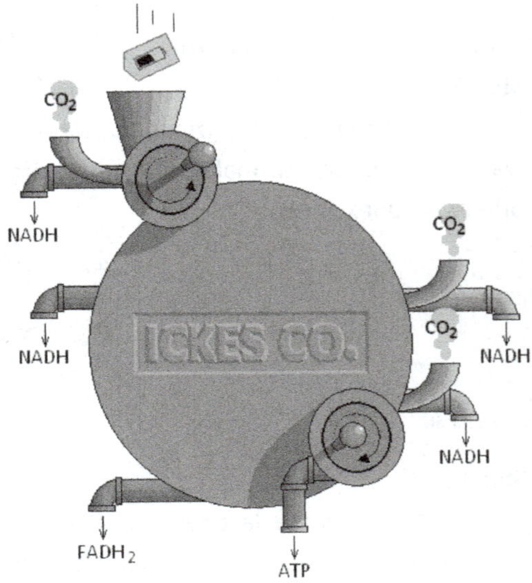

In addition to the freed carbon dioxide, each round of the citric acid cycle (along with the introductory step) makes one ATP (converted from GTP), four NADH and one $FADH_2$. Some of these are produced after all the carbon dioxide has been freed. These energy totals get doubled, since two pyruvates are made per glucose[25]. All the NADH and $FADH_2$ that is made then participates in the last bit of reactions that make energy from oxygen's greedy interest in electrons. That story is soon to follow.

To the right is a cartoonish version of the citric acid cycle showing the overall reaction. As it shows, one pyruvate gets crunched up in a machine to make three CO_2, four NADH, one $FADH_2$, and one ATP.

Electron Transport Chain

We are now at the last part of the energy formation process of respiration. This is a set of reactions called the **electron transport chain** that are *very* similar in principle to the process taking place in chloroplasts. To see how this is so, have a look at the next image.

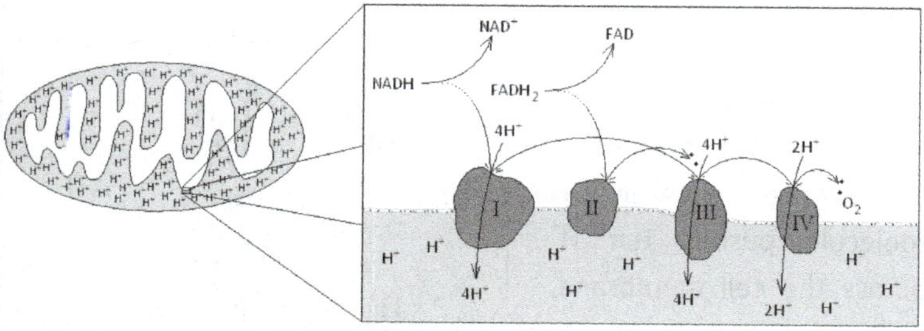

The electron transport chain involves a series of proteins that are imbedded in the inner membrane of mitochondria. Between this

---

25 It may seem like I'm pounding the concept of "two pyruvates per glucose" into the ground, but teachers and professors love to trick students with this on exams. It is critically important to understand that once the glucose is broken in half during glycolysis, *everything* that follows, including the citric acid cycle, happens *twice*.

inner membrane and outer membrane is a space, and a LOT of H^+ are jammed within it. Like the process in chloroplasts, the mitochondria create this overcrowding by "bouncing" electrons along membrane protein islands[26]. To fully appreciate how this happens, recall from our discussion on oxidation and reduction that oxygen loves electrons. In science, this is the same thing as saying that *electrons love oxygen.* NADH and $FADH_2$ drop off their electrons onto protein islands, and as the electrons urgently move from protein to protein in attempt to get to oxygen, the mitochondrion uses this intense energy to push H^+ ions across the membrane and into the crowded space. This cruel overcrowding causes the H^+ to want out, *very badly*. To get out, they pay an "energy toll," which generates ATP from ADP and phosphate.

As the next image shows, the O_2 receiving the electrons breaks apart into two O^{2-} ions, and these rapidly recombine with H^+ to create water. (Each O_2 accepts four electrons). At the same time, the energy of H^+ escaping the inner space gets used to create ATP.

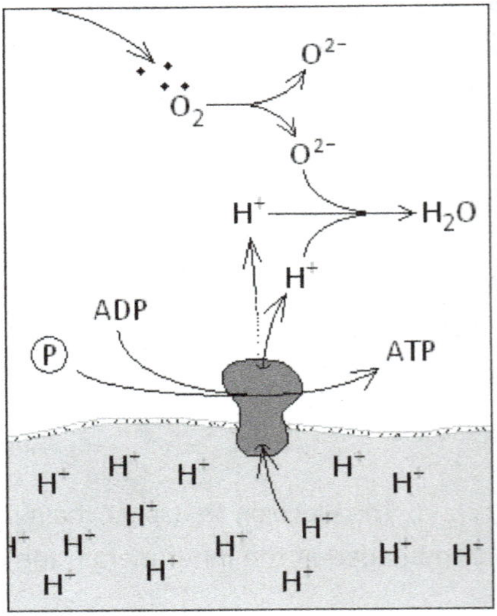

Just how much ATP is made from this process? On average each NADH molecule pushes ten H^+ across the cell membrane, and these protons create three ATP as they escape back out. Each $FADH_2$

---

[26] Once again I must offer a disclaimer: the electrons do not technically "bounce" along the proteins in the membrane, but rather travel through them in a highly energetic fashion. The image of bouncing electrons provides us with a much more memorable idea of what is happening though.

pushes six H^+ across the cell membrane, and these create two ATP as they escape back out. Looking back over glycolysis and the citric acid cycle, we can now begin to take a tally of all the ATP that is created from a single glucose molecule. From glycolysis, we get a net of two ATP plus two NADH. Each NADH is worth three ATP. So, with the use of the electron transport chain (which requires O_2) we get (2 ATP + 2×3 ATP) = 8 ATP total. During the course of the introductory step and subsequent reactions of the citric acid cycle, (which happens twice per glucose), we get another two ATP, another eight NADH, and two $FADH_2$. This amounts to (2 ATP + 8×3 ATP + 2×2 ATP) = 30 ATP total. The grand total ATP harvested from glucose during glycolysis and the citric acid cycle is 8 ATP + 30 ATP = 38 ATP.

Glycolysis	----> 2 ATP + 2 NADH ----> 2 ATP + 6 ATP =	8 ATP
Citric acid cycle	----> 2 ATP + 8 NADH + 2 $FADH_2$ = 2 ATP + 24 ATP + 4 ATP =	30 ATP
		38 ATP

If oxygen is not present, the entire process comes to a screeching halt at the end of glycolysis, yielding only two ATP and two lactic acid molecules. The radical difference in energy is the reason we must breathe oxygen.

The Grand Summary

The formation of water and storage of energy in ATP molecules brings to an end the complex processes of photosynthesis and respiration. The first process stores the energy as glucose, and the second pulls it back out and puts the energy into ATP for the cell to use as general currency. Carbon dioxide and water were used to make the glucose, and as the glucose is broken down, CO_2 and water are released once again. The net result is as follows:

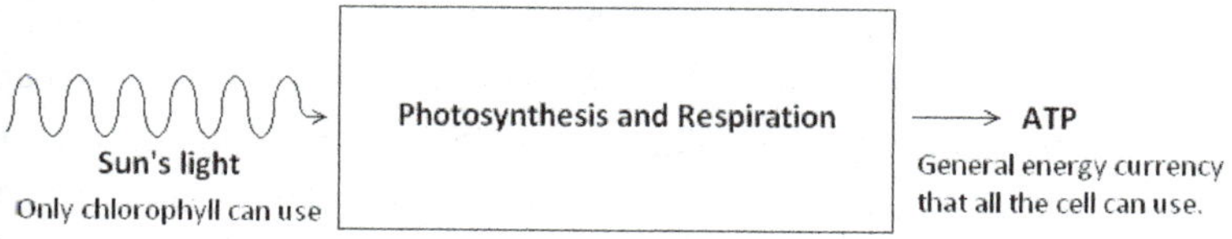

Summary

In this section, we discussed the citric acid cycle and the electron transport chain.

- The two pyruvates that result from glycolysis are each converted to acetyl-CoA, which is the entry point for the citric acid cycle. One CO_2 from each pyruvate is released at this point.
- Pyruvate and acetyl-CoA are chemicals that can be made from the breakdown of amino acids and fatty acids, as well as from glucose. This makes the citric acid cycle and the electron transport chain the common fate of many of the cell's molecules.
- The citric acid cycle creates CO_2, a few ATP and a LOT of electron middlemen. These electron middlemen, (NADH and $FADH_2$), carry electrons to oxygen via the electron transport chain. Electrons want to be with oxygen just as bad as oxygen wants to be with electrons. So, NADH and $FADH_2$ force electrons to do work in making a LOT of ATP in order to get to oxygen.
- Steps two and three of the citric acid cycle remove the remaining two CO_2 from each starting pyruvate. This accounts for all the carbon originally locked in glucose.
- During the course of the citric acid cycle, one GTP is produced, but is rapidly converted to ATP.
- A total of four NADH and one $FADH_2$ are created in the citric acid cycle and its introductory step.
- The electron transport chain is the last step in the energy harvesting process. It takes place in the inner membrane of the mitochondrion. The electrons from NADH and $FADH_2$ are brought to the proteins of the electron transport chain, and these electrons move from protein to protein in a desperate

attempt to get to the oxygen waiting at the end. This energy pumps H^+ across the membrane and builds a large H^+ gradient.

- The H^+ that get pumped across the inner membrane gather in the space between the inner membrane and outer membrane. They are very unhappy in this state and push through a protein canal that extracts their aggravation energy and uses it to make ATP.
- Each NADH makes roughly three ATP. Each $FADH_2$ makes roughly two ATP.
- The grand total energy that can be extracted from glucose is ~38 ATP, assuming oxygen is present to allow the electron transport chain to work. If oxygen is not present, only two ATP and two lactic acids are made.
- By the time this respiration process is finally complete, CO_2 and water are released. This is the reverse of what happens in photosynthesis, in which CO_2 and water are consumed.

Index

D

E

F

R

S

T

U

V

W

X

Made in the USA
Monee, IL
04 November 2026

67377507R00075